Contamination in
SOIL ENVIRONMENT

THE EDITORS

Dr. Ashutosh Gautam obtained his Master Degree from Garhwal University, Srinagar (Garhwal) in 1985 and Doctor of Philosophy from the H.N.B. Garhwal University in 1989.Dr. Gautam is a fellow member of several environmental organizations & academic bodies. He has more than three dozen research papers/ articles & radio talks to his credit. Dr. Gautam has authored/edited more than 12 books related to Environment & Sustainable Development. Dr. Gautam is the editor of an international journal – Environment Conservation Journal & member of editorial board of several international journals – Journal of Environment & Pollution, Indian Journal of Environmental Sciences & Indian Journal of Environment & Eco-Planning. Dr. Gautam was awarded 11th JEB prize (a young scientist award) by the Academy of Environmental Biology, India. Dr. Gautam also received Bharat Jyoti Award from India International friendship society, India in year 2001 and several other awards from various other Academic Societies and NGOs.

Dr. Chakresh Pathak is a young researcher working in the field of Environmental Chemistry, Environmental Pollution, Environmental Impact Assessment & Environmental Toxicology. He earned M.Sc. (2007) & Ph.D. (2012) from Gurukula Kangri University, Haridwar. He has received research fellowship under the scheme of BSR (UGC) for meritorious Research Students. He has authored/edited three books. He has published more then a dozen research paper in National/International Journals. Attended more than Fifteen National and International conferences. He is on the Advisory board of the Journal of Applied and Natural Science & BIOINFO Environment and Pollution. He is fellow member of Applied and Natural Science Foundation. Dr. Pathak was awarded special Appreciation award in Ist WCMANU-2011.

Contamination in SOIL ENVIRONMENT

– Editor –

Dr. Ashutosh Gautam
Dr. Chakresh Pathak

2018
Daya Publishing House®
A Division of
Astral International Pvt. Ltd.
New Delhi – 110 002

Published by : **Daya Publishing House**®
A Division of
Astral International Pvt. Ltd.
– ISO 9001:2015 Certified Company –
4736/23, Ansari Road, Darya Ganj
New Delhi-110 002
Ph. 011-43549197, 23278134
E-mail: info@astralint.com
Website: www.astralint.com

Digitally Printed at : Replika Press Pvt. Ltd.

Preface

The main purpose of the book is to present comprehensive and concise knowledge of the recent advancement in the field of soil contamination & remedial measures. Soil protection has become a major issue and a crucial factor for future technological progress that must meet certain requirements for sustainable development. Chapters are focusing on contamination of soil environment and its impact on human through different type of exposures. Overall the information compiled in this book will bring in depth knowledge and recent advancement in the researches on the metallic contamination of soil & remedial measures.

In the first chapter Farzana Khan Perveen discusses on the Effects of Contaminated Environment on Human, Animals and Plants. In the second chapter Poonam Gusain and team discusses the impact of most toxic heavy metals arsenic (As), cadmium (Cd) and lead (Pb) on different plant species by evaluating changes in morphological physiological and yield characteristics. Veethika Tilwankar, Swapnil Rai and S. P. Bajpai in the third chapter discusses the accumulation of heavy metals in soil and its interaction and toxic effects on human, plant and animal life. Vinod Kumar, A. K. Chopra, Roushan K. Thakur and Jogendra Singh in fourth chapter describe the possible sources, chemistry, potential biohazards and best available remedial strategies for various heavy metals.

In the fifth chapter Vikas Kumar and his team discusses on the bioremediation methods such as biosorption, metal-microbe interactions, bioaccumulation, biomineralisation, biotransformation and bioleaching. Payal Garg and Geetanjali Kaushik in the sixth chapter discusses on the types of earthworms and their role in maintaining soil fertility, vermicomposting, soil erosion control and land reclamation.

The magnitude of heavy metal contamination of soil and ground water in and around unauthorized E-waste recycling site has been studied by Sirajuddin Ahmed, Rashmi Makkar Panwar and Anubhav Sharma. In the Eigth chapter M.K. Jhariya and D.K. Yadav discusses on complex, diverse & dynamics of Riparian areas ecosystems.

Ninth chapter written by Meenu Srivastava and Vinita Koka discusses the Impact of Textile Effluent on Soil Fertility. In Tenth Chapter Preeti Sharma and her team discusses the types, sources and forms of soil contamination.

The use of biomarkers offer a sensitive fundamental approach in the assessment of ecosystem health as it not only evaluates the presence of metals, but allows the early detection of metal induced biological changes. The study carried out by D.D. Kadam on the Genetic Variability, Correlation and Path Analysis in Finger Millet. Chander and Sharma in his chapter provide a brief review on occurrence and remediation of pharmaceutical compounds in soil and sediments.

The present book is an endeavor to amalgamate the researches on Soil contamination & Remediation techniques established by various research scientists.

Editors

Contents

Effects of Contaminated Environment on Human Being, Animals and Plants

Farzana Khan Perveen

Department of Zoology, Shaheed Benazir Bhutto University (SBBU), Main Campus, Sheringal, Dir Upper (DU), Khyber Pakhtunkhwa (KP), Pakistan

Abstract

All sorts of human activities and sometime natural hazards contaminate the environment, air, water and soil. All type of chemicals including heavy metals, natural toxins, pesticides, minerals, oil, detergents, radiations etc are involved in it. Contaminated or polluted soil environment directly affects human health through direct contact by decrease the health quality. Contaminates have a tremendous impact on the ecosystem. They negatively affect the human, animals, birds and plants, externally by altering structure and internally by interfering in metabolism, results sometimes death, however, extreme effect may extinct the species. The intimate linkages between animals and bird species and their habitats make them useful for identifying ecosystem health. As such, they can be indicators of deteriorating habitat quality and environmental pollution. Combined with their ability to signal the eminent outbreak of diseases, they are incredibly useful as indicators to both the fields of environmental science and to human health. Naturalists preserved animals and birds for studies; they can provide perception into past environmental contamination, offering a model to which it can be compared current conditions. The present study aims to educate the public and creates awareness to utilized animals and birds to enhance quality of ecosystem for better survival of all living organisms. Remediation of contaminated environment is necessary to reduce the associated risks, make the land resource available for agricultural production and enhance food security.

Keywords: Contamination, effects, environment, human being, plants.

Environment Contamination

Pollution or contamination may be defined as an undesirable change in the physical, chemical and biological characteristics of air, water (H_2O) and soil which affect human life, lives of animals, plants, industrial progress, living conditions and cultural assets. A pollutant adversely interferes with health, comfort, property or environments of the people, animals and plants. Generally most pollutants are introduced in the environments by sewage, waste, accidental discharge or else they are byproducts or residues from the production of something useful. Due to this our precious natural resources like air, water and soil are getting polluted (Ashraf *et al.,* 2014). Pollutants are generated by all sorts of human activities and sometime by natural hazards. They have to be bioavailable to cause harm. Soil pollution is a wide spread problem and as part of land degradation is caused by the presence of xenobiotic (human-made) chemicals or other alteration in the natural soil environment.

The causes of contamination of environments may be: the industrial activity, direct discharge of industrial wastes to the soil, improper disposal of waste, drainage of contaminated surface water into the soil, agents of war, nuclear wastes, ammunitions, radiation, electronic waste, oil drilling, oil and fuel dumping, mining, landfill and illegal dumping, coal ash, activities by other heavy industries, accidental spills as may happen during activities, corrosion of underground storage tanks (including piping used to transmit the contents), acid rain (in turn caused by air pollution), transportation, trading activities, residences, burning fossil fuels, intensive farming, deforestation, pollution due to urbanization, road debris, genetically modified plants (GMP), dumping of solid wastes, plants and dead animals (contaminated biomaterials), agrochemicals, specially pesticides, insecticides, fungicides, fertilizers etc (Snyder, 2005). The agrochemicals are persisted for various periods of time in the environments (Ashraf *et al.,* 2014).

The most common chemicals involved are petroleum, hydrocarbons, polynuclear aromatic hydrocarbons such as naphthalene and benzo(a)pyrene, polychlorinated biphenyls (PCBs), chlorinated aromatic compounds, detergents, and pesticides. Inorganic species include nitrates (NO_3), phosphates (PO_4), and heavy metals such as arsenic (As), cadmium (Cd), chromium (Cr), copper (Cu), lead (Pb), mercury (Hg), nickel (Ni)and zinc (Zn), inorganic acids and radionuclides (radioactive substances) and solvents etc. Soil also naturally contributes to air pollution by releasing volatile compounds into the atmosphere. Nitrogen (N) escapes through ammonia (NH_3) by volatilization and denitrification. The decomposition of organic materials in soil can release sulfur dioxide (SO_2) and other sulfur (S) compounds, causing acid rain. Heavy metals and other potentially toxic elements are the most serious soil pollutants in sewage (Adriano *et al.,* 1999). Contamination is correlated with the degree of industrialization and intensity of chemical usage. The concern over soil contamination stems primarily from health risks, from direct contact with the contaminated soil, vapors from the contaminants, and from secondary contamination of water (H_2O) supplies within and underlying the soil. Mapping of contaminated soil sites and the resulting cleanup are time consuming and expensive tasks, requiring extensive amounts of geology, hydrology, chemistry, computer modeling skills, and geographic information system (GIS) in environmental contamination, as well as the history of industrial chemistry (Olawoyin *et al.,* 2012).

Effects of Contamination of Environments

Effects on Human Health: Contaminated or polluted soil environment directly affects human health through direct contact with soil or via inhalation of soil contaminants which have vaporized, potentially greater threats are posed by the infiltration of soil contamination into groundwater aquifers used for human consumption, sometimes in areas apparently far removed from any apparent source of above ground contamination. This tends to result in the development of pollution-related diseases. Health consequences from exposure to soil contamination vary greatly depending on pollutant type, pathway of attack and vulnerability of the exposed population. It causes not only physical disabilities but also psychological and behavioral disorders in people. Chronic exposure to Cr, Pb and other metals, petroleum, solvents, and many pesticide and herbicide formulations can be carcinogenic, can cause congenital disorders or can cause other chronic health conditions. Industrial or man-made concentrations of naturally occurring substances, such as NO_3 and NH_3 associated with livestock manure from agricultural operations, have also been identified as health hazards in soil and groundwater (Hogan *et al.*, 1973).

Chronic exposure to benzene at sufficient concentrations is known to be associated with higher incidence of leukemia. Mercury (Hg) and cyclodienes are known to induce higher incidences of kidney damage and some irreversible diseases. PCBs and cyclodienes are linked to liver toxicity. Carbamates and OP can induce a chain of responses leading to neuromuscular blockage. Many chlorinated solvents induce liver changes, kidney changes and depression of the central nervous system. There is an entire spectrum of further health effects such as headache, nausea, fatigue, eye irritation and skin rash for the above cited and other chemicals. At sufficient dosages a large number of soil contaminants can cause death by exposure via direct contact, inhalation or ingestion of contaminants in groundwater contaminated through soil. Radiation and the free radicals produced can damage DNA by causing several different types of lesions, e.g., single strand breaks, double strand breaks, base changes, interstrand crosslinks. Oil exploitation can lead to contamination of environments, which can have effects on human health, however, the relationship between oil contamination and health effects such as cancer incidence, pulmonary ailments, and psychological, reproductive, and dermatologic health problems (Gupta *et al.*, 1982).

Air contamination may affect human by causing reduced lung functioning, irritation of eyes, nose, mouth and throat, asthma attacks, respiratory symptoms such as coughing and wheezing, increased respiratory disease such as bronchitis, reduced energy levels, headaches and dizziness, disruption of endocrine, reproductive and immune systems, neurobehavioral disorders, cardiovascular problems, cancer, premature death (Kati *et al.*, 2004). Waterborne diseases (WBD) caused by polluted drinking water such as typhoid, amoebiasis, giardiasis, ascariasis, hookworm disease. Polluted beach water caused such as rashes, ear ache, pink eye, respiratory infections, hepatitis, encephalitis, gastroenteritis, diarrhoea, vomiting and stomach aches. Ailment conditions related to H_2O polluted by chemicals (such as pesticides, hydrocarbons, persistent organic pollutants, heavy metals etc) are cancer, including prostate cancer, non-hodgkin's lymphoma, hormonal problems that can disrupt reproductive and developmental processes, damage to the nervous system, liver and kidney damage, damage to the DNA. Exposure to Hg in the womb may cause neurological problems including slower reflexes, learning deficits, delayed or incomplete mental development, autism and brain damage, however, in adults, it causes parkinson's disease, multiple sclerosis, alzheimer's disease, heart disease and even death and other effects. Water pollution may also results from interactions between water and contaminated soil, as well as from deposition of air contaminants (such as acid rain), damage to people may be caused by fish foods coming from polluted water (a well-known example is high Hg levels in fish) damage to people may be caused by vegetable crops grown/washed with polluted H_2O (Baetjer, 1983).

Soil pollution causes cancer and leukaemia, however, lead in soil is especially hazardous for young children causing developmental damage to the brain, Hg can increase the risk of kidney damage, cyclodienes can lead to liver toxicity, causes neuro-muscular blockage as well as depression of the central nervous system, also causes headaches, nausea, fatigue, eye irritation and skin rash. Moreover, contact with contaminated soil may be direct (by using parks, schools etc) or indirect(by inhaling soil contaminants, which are in vaporized form). Further, soil pollution may also results secondary contamination of H_2O supplies and deposition in air contaminants (for example, via acid rain), contamination of crops grown in polluted soil brings up problems with food security. Since it is closely linked with H_2O pollution, many effects of soil contamination appear to be similar as caused by H_2O contamination (Debackere, 1983) (Figure 1). Air and H_2O pollution can have negative impacts on human health, but the impacts of such soil pollution on our health have had a much lower profile, and are not well understood.

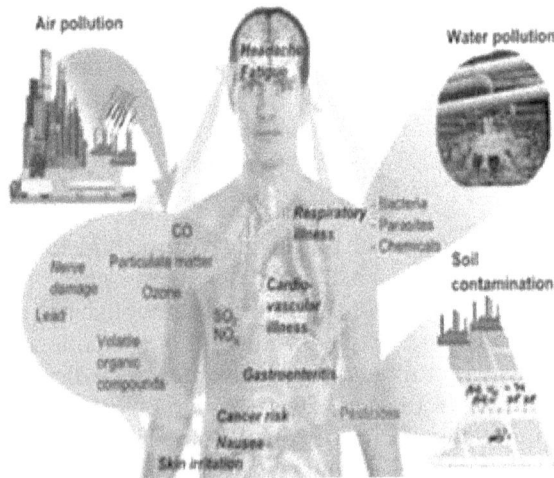

Fig. 1: Effects of contamination of environments on human health causing different diseases in overall body (Donia *et al.*, 1974)

Effects on ecosystem: Pollutants like oil, detergents, N and P from fertilizers and Pb can have a tremendous impact on the ecosystem, especially if the H_2O gets polluted. In a lake, for example, it can wreak havoc on the ecological balance by stimulating plants growth and causing the death of fish due to suffocation resulting from lack of the oxygen (O_2). Its cycle will stop, and polluted H_2O will also affect the animals dependent on the lake H_2O. Not unexpectedly, soil contaminants can have significant deleterious consequences for ecosystems. There are radical soil chemistry changes, which can arise from the presence of many hazardous chemicals even at low concentration of the contaminant elements. These changes can manifest in the alteration of metabolism of endemic microorganisms and arthropods resident in a given soil environment. The result can be virtual eradication of some of the primary food chain, which in turn could have major consequences for predator or consumer species. Even if the chemical effect on lower life forms is small, the lower pyramid levels of the food chain may ingest alien chemicals, which normally become more concentrated for each consuming rung of the food chain (Ruckebusch *et al.*, 1983). Many of these effects are now well known, such as the concentration of persistent dichlorodiphenyl trichloroethane (DDT) for avian consumers, leading to weakening of eggshells, increased chick mortality and potential extinction of species. Effects occur to agricultural lands which have certain types of soil contamination. They typically alter plant metabolism, often causing a reduction in crop yields. This has a secondary effect upon soil conservation, since the languishing crops cannot shield the earth's soil from erosion. Some of these chemical contaminants have long half-lives and in other cases derivative chemicals are formed from decay of primary soil contaminants such as habitat destruction and land degradation (Agarwal *et al.*, 2016). Environmental pollution, almost exclusively created by human activities, has a negative effect on the ecosystem, destroying crucial layers of it and causing an even more negative effect on the upper layers.

Effects on Animals: Domestic animals are kept in many different climates, from the harsh-cold environments of the north to the hot-arid environment of the deserts and the hot-wet climates of the tropics. Stressful climatic conditions, either hot or cold, tend to aggravate the impact of xenobiotics on livestock production. The problem is further aggravated when modern intensive agricultural techniques, with increased stocking densities, place many animals in close proximity on a single site. Conversely, animals that are distributed over large rangelands are far less subject to the dangers of massive chemical exposures. Chemical substances can enter the environment of domestic animals by direct application or in some instances by complex pathways. Application of pesticides to animals or their housing facilities, the use of fertilizers, including animal wastes or herbicides or insecticides to cropland, and the use of sanitizing agents or veterinary drugs are examples of the direct exposure route, whereas indirect entrance may result from food-chain contamination, industrial wastes and accidentsor from combustion processes. The occurrence of chemical residues in domestic animal meat and by-products, including eggs and milk, reflects the increase in the use of agricultural chemicals as well as an increase in pollution of the environment in which the animals are reared and processed. Tropical, arid, and sub-polar regions may alter the utilization of these chemicals in animal agriculture. This is particularly true of veterinary drugs used in microbial infections, in the prevention of disease and infections, and in parasitic control or treatment. Phytotoxins are another case because of the impact that environmental conditions have on their diversity (Walker, 1983). The oil that leaked into the ocean from the Exxon-Valdez had a drastic effect on the life and ecosystems of the area. Mammals and seabirds were affected significantly due to their frequent need to be at the surface of the ocean. It is estimated that between 1000-2800 sea otters, *Enhydra lutris* Linnaeus 1758 and 250,000 sea birds died (Ballachey *et al.*, 2003). Sea birds frequented the surface of the ocean in search of fish but were often immobilized by the layer of oil. Mammals such as sea otters, which surface to breathe, inhaled the toxic oil. The oil spill also had fatal effects on the herring and salmon fishery populations of the area (Picou, 2009).

Toxicoses in Animals

Toxicants, natural and synthetic, may impact on animal agriculture in different ways: by directly or indirectly intoxicating animals (toxicoses), resulting in mortality or decreased production of edible food products; by decreasing the availability or usability of nutritious feedstuffs due to the presence of naturally occurring toxins or added toxicants; and by decreasing the wholesomeness of edible food products due to the presence of hazardous residues (Shull and Cheeke, 1983). There is a wealth of information on domestic animal toxicology in the temperate zone (Osweiler *et al.*,1985). However, in the arid, tropical or sub-polar regions, the light, temperature (both cold and hot), and rainfall are effected on these domestic animals. They can maintain a thermo neutral zone over a range of ambient temperatures. When the ambient temperature moves outside the thermo neutral zone, the animal must alter its metabolic rate to maintain a constant core body temperature. Animals reared outside the temperate region are subjected to large swings in temperatures beyond the thermo neutral zone. Animal metabolism and behaviour are altered, characterized by a shift in food and H_2O consumption, passage rate of food through the digestive tract, hormone synthesis and release, panting and shivering, and reduced metabolic activity. With an altered metabolism there is a potential for xenobiotics to be handled differently by animals (Yousef, 1985a). The thermoregulation can be affected by a variety of toxic substances, thus altering the ability of domestic animals to tolerate thermal shifts in ambient temperature. For example, it has been demonstrated that the organophosphates (OP), parathion or chlorpyrifos reduce the tolerance of animals to cold exposure (Ahdaya *et al.*,1976); and birds (Rattner *et al.*,1982). Carbamate and OP pesticides are known to be cholinesterase inhibitors and to cause hypothermia in animals. Acetylcholine is one of the hormones involved in maintenance of body temperature, thus, cholinesterase inhibitors have an effect on thermoregulation in mammals (Ahdaya *et al.*,1976) and birds (Rattner *et al.*,1982). The inability to thermo regulate may not be the cause of greater death losses, at least in birds, rather, the cause may be reduced insulation brought about by loss of subcutaneous fat or depletion of carbohydrate and lipid reserves induced by decreased food intake (Rattner *et al.,* 1982). Species that are indigenous to a region are better able to maintain a more constant core body temperature than non-adapted animals. This ability is brought about through anatomical and physiological adaptations that have occurred. The question has been asked, do arctic birds and mammals maintain body temperatures within the same range as species from temperate regions? Both birds and mammals subjected to environmental temperatures of -30 °C and -50 °C maintained their body temperatures within normal limits (Nielsen, 1979). Protected by insulation, animals in arctic climates do not need to eat more than animals in milder climates (Yousef, 1985b). The climate of the arctic region is not always cold, but rather is characterized by extreme variation from winter cold to summer heat and great differences in sunlight between winter and summer days (Irving, 1964). Non-adapted animals experience difficulty in adapting to these extremes. In general, animals indigenous to tropical areas of the world are better adapted for heat exposure than those from temperate regions (Ingram and Mount, 1975; Yousef, 1982).

In the tropical areas, animals such as cattle possess an increased ability to lose heat due to a greater surface area in the region of the dewlap and prepuce and increased numbers of sweat glands and the presence of short hair. Fat stores may be in humps or inter muscular rather than subcutaneous, which assists the conductance of heat from the core to the surface skin. Animals from the temperate region with high productivity have often failed to continue to yield as well when exposed to tropical climates with extremes in temperature. In hot and arid climates adaptation mechanisms aid in the maintenance of normal body temperatures. Heat storage, insulation, panting, gular fluttering, and blood flow mechanisms to cool the brain are just some of the physiological adaptations used by animals in hot climates with limited H_2O supplies (Nielsen, 1979). Because domestic animals are homeotherms and indigenous animals are adapted to withstand climatic variations, the problems with xenobiotic exposure are not greatly different from those in the temperate region. However, several problems are amplified. At high ambient temperatures, there is an increase in water consumption and a concomitant decrease in feed consumption (National Research Council, 1981). If exposure to a chemical is via H_2O the exposure will be increased and the animal may be at greater risk. In cold climates, feed consumption increases during extremes of cold making exposure via the food chain an increased problem. It is known that food restriction and/ or water deprivation may significantly alter the response of an animal to toxic chemicals (Baetjer, 1983). These changes in food and H_2O consumption, which mark the principal metabolic shift in animals in response to environmental fluctuations, may contribute to toxicological differences in pesticides between the world's regions.

Natural Venoms

Toxigenic fungi have ubiquitous geographical distribution influenced by climatic conditions, cultivation and harvesting techniques, as well as storage procedures and the livestock production practices used. Mycotoxins occur in particular feeds and in particular regions (National Research Council, 1979; Smith, 1982). Aflatoxins are comparatively common in subtropical regions and depend on factors such as weak plants resulting from drought stress, insect or mechanical damage, climatic conditions before drying and improper storage conditions (Galtier and LeBars, 1983). In tropical areas of the world, mycotoxins in grain, protein concentrates and other feedstuffs are a major problem because warm and humid environmental conditions favour fungal growth and farming practices in many tropical areas are not sophisticated (Cheeke and Shull, 1985). Also, crop storage conditions are frequently inadequate in these areas. Thus, in hothumid regions the production of natural chemicals in feedstuffs may result in toxicosis, which poses a problem for livestock production.

Incidence of Pesticides

Chemical accidents that adversely affect animal agriculture have occurred in the past and will doubtless occur in the future. These problems are not unique to any area of the planet and do not hinge on climatic conditions. As an example, when organochlorine (OC) insecticides were being phased out because of their adverse effects on non-target organisms, together with their persistency in the environment and their carcinogenicity, OP insecticides were introduced as logical replacements. Some of these compounds were halogenated phenyl phosphonates and phosphonothionates that were lipid soluble, persistent and lower toxicity to mammals than the parathions and other widely used OP insecticides. However, a number of them were known to be delayed neurotoxins (Metcalf, 1982). Leptophos, one such chemical, was used in 1971 to control the cotton leafworm, *Spodoptera littoralis* (Boisduval) in Egypt. Some 1300 *Boss indicus* Linnaeus 1758 had been suffered paralysis and distal axonopathy characteristic of delayed neurotoxicity (Donia *et al.*, 1974). Human poisoning was also evident (Hassan *et al.*, 1978). Egypt was only one of about 50 countries into which leptophos was sold. This example of toxicosis in livestock is one, where H_2O from cropland collected in a river and H_2O consumption in a hot and arid climate caused the death of many animals. This is not an isolated case of chemical toxicosis in livestock, but documentation in the literature is not common. In temperate regions there are numerous reports of poisoning in intensified animal production units (Shull and Cheeke, 1983). These accidents have included such chemicals as polychlorinated biphenyls, polybrominated biphenyls, tetrachlorodibenzo-pdioxins, and OC insecticides.

Mineral as Contaminates

Lead is considered to be one of the major environmental pollutants and has been incriminated as a cause of accidental poisoning in domestic animals in more cases than any other substance (National Research Council, 1972). It contaminates the environment is largely air-borne but is redeposited by dust into soil and H_2O, however, it is taken up by or exists on the surface of plants, which are grazed by livestock. *B. indicus*; sheep, *Ovis aries* Linnaeus and horses, *Equus ferus* Linnaeus are good indicators of pollution on vegetation (Debackere, 1983). Toxicosis of Pb in cattle from the use of Pb-based pigments in paint was common, however, it was poisoning of waterfowl, *Gallus domesticus* Linnaeus by spent Pb-shot. Restriction in the use of Pb-based paints and, currently, in Pb-shot has reduced the problem in the United States of America (USA). Lead from smelters may cause problems in *E. ferus* grazing in adjacent areas. Dog, *Canis domesticus* Linnaeus, 1758 and cat, *Felis domesticus* Linnaeus, 1758 give a very good indication of Pb pollution in urban areas as the concentration of Pb in their livers and kidneys increases with increased pollution (Debackere, 1983). Animals can be exposed to Hg contamination from air, soil, H_2O and ingestion of contaminated feed. Contamination of Hg results from fossil fuel combustion, agricultural fungicides, smelting of commercial ores, and through industrial discharge, contaminates H_2O and, then ingestion by fish through H_2O. The fish are incorporated into animal feeds by way of fish meals or protein concentrates. Toxicosis of domestic animals has also been due to the consumption of contaminated grain. In 1971, Iraqi authorities ordered 73,000 tons of wheat, *Triticum aestivum* Linnaeus and 22000 tons of barley, *Hordeum vulgare* (Linnaeus) from suppliers in Mexico and Canada, respectively that were treated with Hg. This grain was used for planting but some was prepared into homemade bread. Oral ingestion may have included meat and other animal products obtained from the livestock given the treated grain. The latent period between dose and onset of symptoms may have given farmers a false sense of security since *G. domesticus* given *T. aestivum* for a period of a few days did not die (Bakir *et at.*, 1973) but 6530 cases were admitted in Hospital (Clarkson *et at.*, 1976). Both man and animals are subject to Hg toxicosis through contaminated grains. In Minamata Bay in Japan, methyl Hg poisoning was observed in *F. domesticus* before human cases were recognized. The common thread between the disease in *F. domesticus* and that in humans was shown to be the consumption of Hg contaminated fish (Hodges, 1976). Under modern feeding and management conditions of livestock production, Cd toxicosis is relatively unimportant (Neathery and Miller, 1975), but it does not preclude ingestion of recycled waste material, such as sewage sludge, in which Cd may be concentrated. Cadmium is toxic and is an antagonist of Zn, Fe, Cu and other elements. Some plants, such as clover, *Trifolium acaule* Rich have capacity to concentrate Cd from soil. Arsenic (As) may result in contamination of livestock in areas surrounding smelters and where As compound are used for weeds and insects control. Since fish are often high in As, fish meals may contribute sizeable quantities of As to livestock. Non-ruminants are generally more susceptible to intoxication than are ruminants or *E. ferus*. The degree of toxicity in ruminants is variable and may depend on the route of exposure, age, nutritional status, and exposure duration (Case, 1974). There are relatively few veterinary examples of acute Cu toxicosis except in cases of accidental overdosing or the consumption of Cu-containing compounds. *Ovis aries* are extremely sensitive to excess Cu and, therefore, are good indicators of environmental pollution by Cu. Concentrations two- to three-fold above normal grass Cu concentrations of 8-15 ppm are toxic to *O. aries* (Debackere, 1983). Selenium (Se) is used in some areas as a supplement to animal diets, still in other areas; there may be Se toxicity due to high levels. Dietary Se requirements are approximately 0.1-0.3 ppm, while toxic concentrations are about 10-50 times greater (National Research Council, 1980). When pasture is limited in dry weather, accumulator plants may be readily available and eaten by livestock, resulting in Se poisoning. In the USA, irrigation of arid land with H_2O contains high concentration of certain minerals, including Se has resulted in a large site where *G. domesticus* have shown evidence of what appears to be Se toxicosis. There some adverse effects of minerals have been found in livestock production. In some cases, climatic conditions alter their consumption but in most instances, exposure is indirect (Selby *et at.*, 1974).

Environmental Indicators

A variety of instruments have been invented to monitor the health of ecosystems. For example, to examine H_2O quality in a wetland, an environmental scientist may use a sensor to measure dissolved O_2 in the H_2O or perform chemical assays in the lab to examine heavy metals in the soil. However, in some cases changes are examined in the habitat without instruments, by studying the behavior of animals and birds, which can inform us about changing ecosystems. Any living organism that is used in such a manner to measure environmental conditions is called an environmental indicator species. Perhaps the best known example is the proverbial canary, *Serinus canaria* (Linnaeus, 1758) in a coal mine. Since *S. canaria* are very susceptible to poisonous gases like carbon monoxide (CO), carbon dioxide (CO_2) and methane (CH_4). They are affected by poisonous gases beforehumans; therefore, miners would take *S. canaria* into the coal mines when they went to work. If *S. canaria*began to show signs of poisoning, it would give miners a chance to put on a mask or escape from the mines before they too succumbed to poisonous gases. The use of birds and animals to monitor environmental conditions continues because birds and animals can tells us a suite of useful information about the environment (Butler *et al.*, 2012).

It is clear that domestic animals can serve as indicators for environmental pollution by chemicals. The role of domestic animals as indicators for environmental pollution through pesticides is negligible (Debackere, 1983). Based on present evidence, fish and birds appear more susceptible than mammals to pesticides, especially OP, carbamates, and chlorinated hydrocarbons (Walker, 1983). Physiological and anatomical differences are likely to affect susceptibility to a wide range of compounds. In domestic birds, the excretory route of these elements is via egg laying and the fact that blood from the gut goes to the kidney via the renal portal system prior to hepatic contact conveys an advantage to birds over mammals, however, the high body temperature, urinary release into the cloaca, and relatively small liver render birds more susceptible to pesticides (Debackere, 1983). Physiological and biochemical evidences suggest that birds have less effective defense mechanisms than do mammals to xenobiotics. Generally, indicator species are chosen for their toxicological susceptibility (Kenaga, 1978). *Gallus domesticus* are regularly used in the laboratory as predictive models for delayed neurotoxicity by OP chemicals. In addition to *G. domesticus*; human, *Homo sapiens* Linnaeus, 1758; water buffalo, *Bubalus bubalis* (Linnaeus, 1758); *E. ferus*; *B.indicus*; *O. aries*; pig, *Sus domesticus* Erxleben, 1777; *C.domesticus* and *F. domesticus* have been reported to be sensitive compared to other common laboratory animals, such as the rat, *Rattus ranjiniae* Agrawal and Ghosal, 1969; mouse, *Rattus norvegicus* (Berkenhout, 1769); rabbit, *Oryctolagus cuniculus* (Linnaeus, 1758); guinea pig, *Cavia porcellus* (Linnaeus, 1758); hamster, *Phodopus sungorus* (Pallas, 1773); and gerbil *Gerbillurus paeba* (Smith, 1836) are not. The adult *G. domesticus* is utilized most frequently as the test animal, however, *F. domesticus* can be used and may serve as an excellent model in extrapolation to man. Wild birds are invaluable models for environmental toxicology due to their abundance, visibility, and diverse habitat associations (Hill and Hoffman, 1984), and are used to monitor pollution in urban as well as in aquatic environments. In Guatemala, hog/ boar, *Sus scrofa* Wagner, 1839 fed on *T. aestivum* seed treated with organo-Hg as a fungicide developed blindness, lack of coordination, and posterior paralysis in 2-3 weeks (Ordonez *et al.*,1966). Humans followed with similar signs of toxicosis from eating the same *T.aestivum* seed. When it comes to chemicals introduced into our environment by indirect pathways, as referred to earlier, domestic birds as well as wild ones, *O. cuniculus, F. domesticus,* and *B. indicus* have been biological indicators for the presence of contamination.

Good Environmental Indicators

When birds come to usefulness as an indicator, all species are not created equal and a few criteria are required for a species to be valuable in this regard. Firstly, it should be sensitive to changes in the environment in order to serve as an early warning. A species that is extraordinarily resilient and not dramatically impacted by environmental changes would offer little information about what is happening in the environment. Additionally, the species needs to respond to change in a predictable manner. If it responds erratically to change, this would make it hard to interpret the underlying environmental causes of the changes that are observed. Lastly, it should be easy to compile and interpret data on the species to inform policy decisions. Species that are very rare would make poor indicator species, because it would be hard to find and study enough of them to draw any meaningful conclusions. Similarly, it would be difficult to gather data on species that have very cryptic life histories or that are in general poorly understood, making them less than ideal candidates for indicator species. By establishing this criterion, the different kinds of ecosystem changes may be explored that birds can tell us about them (Stolen *et al.*, 2005).

Changes in Habitat

One of the most useful things that birds can indicate is overall habitat quality. When birds are dependent on the habitat functioning in specific ways, the population trends of birds can tell us about how well the ecosystem functions. For example, numbers of wading birds (Birds that commonly occur in reedy areas, shallow waters, ponds), such American flamingo, *Phoenicopterus ruber* (Linnaeus, 1758) nesting in the Everglades are dependent on prey availability. The construction of canals and levees to alter the flow of H_2O in the Everglades in the 1950s severely degraded the ecosystem. This led to less prey availability, which caused massive declines in the annual number of wading birds nesting in the Everglades. These declines began before anyone realized that significant damage had been done to the Everglades ecosystem by human activities and *P. ruber* responded well before any other animals. The sensitivity of these birds to proper ecosystem functioning makes them valuable indicators of habitat quality (Stolen *et al.*, 2005).

In some cases, it is not just the numbers of birds present, but the assemblage of bird species in an area that can indicate habitat quality. A study in the Central Appalachian Mountains (CAM) showed that when forest habitats became degraded, the types of birds present changed in a predictable fashion. Birds were classified into categories based on behavioral and physiological response guilds and a Bird Community Index Score (BCI) was calculated based on the types of birds present. As habitats shifted from undisturbed to degraded, there was a corresponding shift from specialist to generalist species because disturbed habitats could not support very specialized species. Examples of this included shifts from bark probers to omnivores and from canopy nesters to open ground nesters. The study showed that calculating the BCI in a particular habitat was a good indicator of its quality (O'Connell *et al.*, 2000).

Similarly, the presence or absence of very specialized species can indicate habitat quality. For example, the red-cockaded woodpecker, *Leuconotopicus borealis* Vieillot, 1809 has very specific nesting requirements. They require living pine, *Pinus radiate* Don trees of large diameter, typically more than 80 years old, to excavate nest cavities in *P. radiate* trees. In newly planted forests, where trees are of smaller diameters, the birds will not nest. Furthermore, they abandon trees when understory plants reach the height of the nest cavity. The suppression of fire causes large growths in the understory, making the habitat unsuitable for nesting. Proper ecosystem functioning in these *P. radiate* tree forests requires periodic burning of understory growth. Because these birds have such specific nesting requirements, which mandate that the habitat be well-functioning, the numbers of *L. borealis* in *P. radiate* tree forests can be effective indicators of habitat quality (Hogan *et al.*, 1973). Since bird numbers can reflect the quality of the habitat, they can also be used to measure the effectiveness of habitat restoration. This principle has been employed in restoration efforts for the Florida Everglades. Alteration of H_2O flow led to concentrations of water in large ponds, which favored birds like egrets that hunt by sight in deep waters. However, birds like ibises (Australian white), *Threskiornis moluccus* Cuvier, 1829 and storks, *Ephippiorhynchus senegalensis* that use tactile feeding and relied on concentrated prey in shallow areas suffered. Because a goal of Everglades restoration is to restore widespread H_2O flow, which would create more areas of shallow H_2O, the ratio of *T. moluccus* and *E. senegalensis* to egrets is one metric that can be used to gauge the success of restoration efforts (Butler *et al.*, 2012).

Another indicator of restoration is the frequency of large nesting events by white ibises, e.g., *T. moluccus*, which was a defining characteristic of the Everglades prior to drainage. The return of these nesting events would indicate that the ecosystem was functioning properly because a dynamic and cyclical nature is a key trait of wetland ecosystems. These examples illustrate how managers can utilize the concept of birds as indicator species in order to monitor the efficacy of habitat restoration (O'Connell *et al.*, 2000; Stolen *et al.*, 2005).

Contamination

Another common use of birds as indicators pertains to contamination/ pollution. Perhaps the best known example of this is the decline in bird species due to the use of DDT, which was brought to international attention in Rachel Carson's "Silent Spring." Birds were the first group of animals whose populations began to noticeably decline as a result of DDT, signaling negative environmental consequences from the pesticide (Bouwman *et al.*, 2013). Accumulation of DDT in the body often resulted in females laying thin eggshells that were crushed during incubation, greatly decreasing reproductive output. For raising public awareness of environmental concerns, birds are also useful because they are ubiquitous through the most habitats and their absence is conspicuous (Balmford, 2013). Therefore, scientists didn't have to convince people that there were decreases in bird populations from DDT, as they would have had to do for pelagic or very obscure species, because the declines were readily apparent to most people. Additionally, since there is a long history of preserving birds for museum specimens, scientists can examine them to determine environmental contamination in the past. The trend of decreasing eggshell thickness from DDT was confirmed by examining eggs preserved in museums that dated back to the 1880s. A similar analysis at the Swedish Museum of Natural History (SMNH) showed increased concentrations of Hg accumulated in birds beginning in the 1940s and 1950s as a result of human activities (Mikusiński *et al.*, 2001) (Figure 2).Radioactivity is a natural phenomenon. It occurs when overly excited atoms seek stability by tremendously emitting energy in the form of radiation. The range in lethality from acute exposure to radiation various among organisms, with mammals being among the most sensitive and viruses being among the most radio resistant (Whicker and Schultz, 1982)

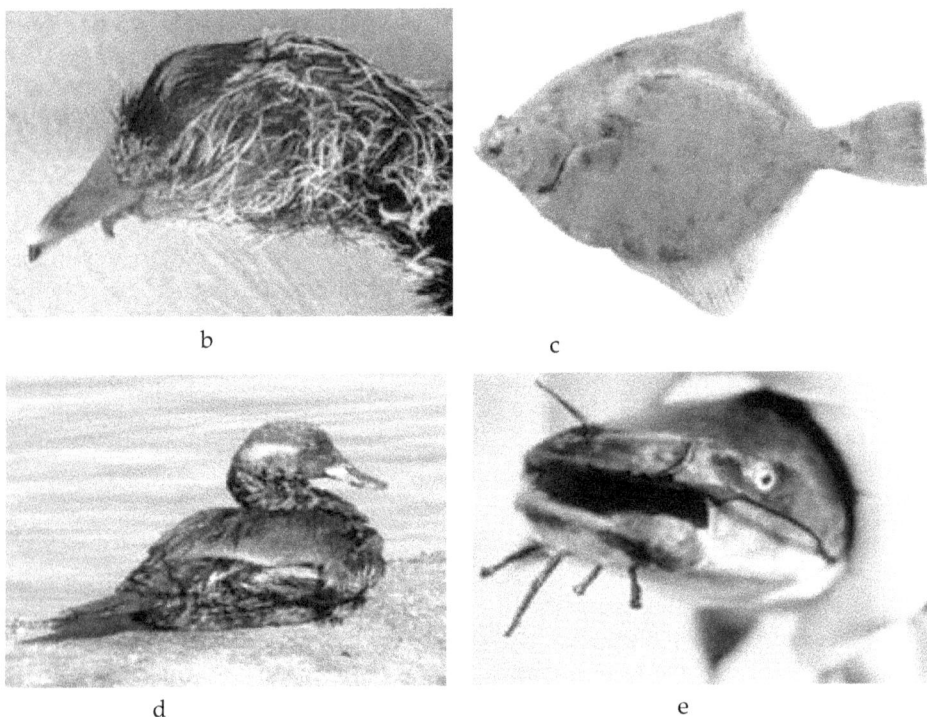

b

c

d

e

Fig. 2: Deformation due to environmental contaminants: a: contaminated area where different type of birds gathered, feed and infect with poisonous elements (Bouwman *et al.*, 2013); b: American coot embryo, *Fulica americana* Gmelin, 1789 from Kesterson National Wildlife Refuge (KNWR), California, USA with selenium-induced developmental abnormalities including a deformed lower bill and no eyes (Baumann *et al.*, 1990);c: exposure to copper concentrations made fish,*Achirus* sp Lacépède, 1802 lose their sense of smell and, therefore, reduce their appetite and food intake (Solomon, 2009); d: Oil coated duck, *Anas platyrhynchos* Linnaeus, 1758 (Donia *et al.*, 1974); e:external neoplasms (tumors) and deformed chin barbells on a brown bullhead catfish, *Ameiurus nebulosus* Lesueur, 1819 due to polycyclic aromatic hydrocarbons (PAH's) from coke ovens, steel-making cities long discharged in the lower Black River, Ohio, USA (Baumann *et al.*,1990)

Indian vulture, *Gyps indicus* (Scopoli, 1786) in Asia have also been sentinels of pollution (Pilastro *et al.*, 1993). Widespread mortality of *G. indicus* indicated that there was a significant problem in the environment. It took more than 10 years after the first documented declines for scientists to figure out the problem, the drug diclofenac. This drug was administered to livestock in order to help them heal from wounds, but it is toxic to *G. indicus*. When *G. indicus* scavenged livestock that had treated with the drug, they ingested it, which caused renal failure and death. Following this, the drug was banned and removed from the market. As in the example of DDT, *G. indicus* showed an early indication of contamination in the environment and provided the impetus for its removal (Blair, 1999). One of the reasons that birds are useful as pollution indicators is that it's relatively easy to collect specimens for detection and it can be done in a noninvasive manner. Feathers dropped by birds contain the heavy metals, as well as they have accumulated in the body, therefore, the amounts are correlated. This means that by analyzing a feather, it is possible to determine the levels of heavy metals inside the body of the bird. By comparing feathers from different populations, scientists can compare pollution in different areas (Jayakumar, 2013). Birds also transfer heavy metals to the shells of the eggs which they lay, providing another way to detect contamination. A study of nonviable eggs from abandoned peafowl, *Pavo cristatus*, Linnaeus, 1758 nests showed that these birds had accumulated levels of Pb and Cd that could cause toxicity. This technique also provided a noninvasive method to determine the presence of contaminants in the environment (Suarez and Tsutsui, 2004). As collecting animal specimens for preservation was common in early naturalists' studies, these birds can provide insight into past environmental contamination, offering a baseline to which it can be compared current conditions.

In the past several decades, amphibian populations have declined across the world, for unexplained reasons which are thought to be varied but of which pesticides may be a part. Pesticide mixtures appear to have a cumulative toxic effect on Indian bull frogs, *Hoplobatrachus tigerinus* Daudin 1803. Tadpoles from ponds containing multiple pesticides take longer to metamorphose and are smaller when they do, decreasing their ability to catch prey and avoid predators. Exposing tadpoles to the organochloride endosulfan at levels likely to be found in habitats near fields sprayed with the chemical kills the tadpoles and causes behavioral and growth abnormalities. The herbicide atrazine can turn male frogs into hermaphrodites, decreasing their ability to reproduce. Both reproductive and non-reproductive effects in aquatic reptiles and amphibians have been reported. Crocodiles, *crocodylus niloticus* Laurenti 1768 many turtle, *Trachemys scripta* (Neuwied, 1839) species and some wall lizards, *Podarcis muralis* Laurenti 1768 lack sex-distinct chromosomes until after fertilization during organogenesis, depending on temperature. Embryonic exposure in *T. scripta* to various PCBs causes a sex reversal. Across the USA and Canada disorders such as decreased hatching success, feminization, skin lesions, and other developmental abnormalities have been reported(Damalas and Eleftherohorinos, 2011).

Pests may evolve to become resistant to pesticides. Many pests will initially be very susceptible to pesticides, but following mutations in their genetic makeup become resistant and survive to reproduce. Resistance is commonly managed through pesticide rotation, which involves alternating among pesticide classes with different modes of action to delay the onset of or mitigate existing pest resistance (Miller, 2004).

Effects on plant

Soil pollution indicates the phenomenon that introducing physiological toxicity substances or an excess of plant nutrition into the soil, the properties of soil degrades or the plant physiological function of soil comes into disorder. Fixation of N_2, which is required for the growth of higher plants, is hindered by pesticides in soil. The insecticides DDT, methyl parathion, and especially pentachlorophenol have been shown to interfere with legume-rhizobium chemical signaling. Reduction of this symbioticchemical is signaling results in reduced N_2 fixation and thus reduced crop yields. Root nodule formation in these plants saves the world economy $10 billion in synthetic N_2 fertilizer every year.Pesticides can kill bees and are strongly implicated in pollinator decline, the loss of species that pollinate plants, including through the mechanism of Colony Collapse Disorder (CCD), in which worker bees from a beehive or western honey bee colony abruptly disappear. Application of pesticides to crops that are in bloom can kill honeybees, which act as pollinators. The USDA and USFWS estimate that US farmers lose at least $200 million a year from reduced crop pollination because pesticides applied to fields eliminate about a fifth of honeybee colonies in the US and harm an additional 15%. On the other side, pesticides have some direct harmful effect on plant including poor root hair development, shoot yellowing and reduced plant growth (Lamberth *et al.*, 2013). Plants, and especially trees, can be destroyed by acid rains; therefore, it will also have a negative effect on animals as well, as their natural environment will be modified. Ozone in the lower atmosphere block the plant respiration and harmful pollutants can be absorbed from the H_2O or soil.

To achieve a healthier soil community, there are a number of important steps to be taken. We need to change how we grow crops to avoid the loss of important microbes and pollinators, and some agricultural practices, e.g., minimum tillage and use of alternative fertilizers must be promoted, such as composts are being tested and used on farms. But we still need to improve our understanding of how soil works; consequently we can design better management systems for greater nutrient availability and faster and stronger plant defense responses when challenged by pests and diseases. We are now well placed to try and tailor our crop plants to select a beneficial community of microbes to 'replace' some chemical inputs.

References

Adriano, D. C., Bollag, J. M., Frankenberger, W. T. and Sims, R. C. eds. (1999). Bioremediationofcontaminated soils. Agronomy monograph 37, American Society of Agronomy, New York, USA 1-25.

Agarwal, A., Zhou, Y. and Liu, A. (2016). Remediation of oil contaminated sand with self-collapsing air microbubbles. *Environ. Sci. Poll. Res.* DOI: 10.1007/s11356-016-7601-5.

Ahdaya, S. M., Shah, P. V. and Guthrie, F. E. (1976). Thermoregulation in mice treated with parathion, carbaryl or DDT. *Toxic. Appl. Pharmaca.* 35: 575-580.

Ashraf, M. A. Maah, M. J. and Yusoff, I. (2014). Soil contamination, risk assessment and remediation In: C. H. S. Maria (Ed), Environmental risk assessment of soil contamination. InTech Publisher, Rajika, Croatia 3-56. DOI: http://dx.doi.org/10.5772/57287.

Baetjer, A. M. (1983). Water deprivation and food restriction on toxicity of parathion and paraoxon. *Archs. Envir. Hlth.* 38 (3): 168-171.

Bakir, F., Damluji, S. F., Amin-Zaki, L., Murtadha, M., Khalidi, A., Al-Rawi, N. Y., Tikriti, S., Dhahir, H. I., Clarkson, T. W., Smith, J. C. and Dohert, R. A. (1973). Methylmercury poisoning in Iraq. *Science* 181: 230-241.

Ballachey, B. E., Bodkin, J. L., Esler, D., Peterson, C. H., Rice, S. D.and Short, J. W. (2003).Longterm ecosystem response to the Exxon-Valdez oil spill. In *Science* 302(5653):2076-2082.

Balmford, A. (2013). Pollution, politics, and vultures.*Science*339 (6120): 653-654.

Baumann, P. C., Harshbarger, J. C. and Hartman, K. J. (1990). Relationship between liver tumors and age in

brown bullhead populations from two Lake Erie tributaries. *Sci. Tot Environ* 94: 71–87.

Blair, R.B. (1999). Birds and butterflies along an urban gradient: surrogate taxa for assessing biodiversity?*Ecol. Appl.*9 (1): 164-170.

Bouwman, H., Bornman, R., Berg, H. V. D. and Kylin, H. (2013). 11 DDT:Fifty years since silent spring:Late lessons from early warnings.*Sci. Precau. Innov.* 240-259.

Butler, S. J., Schel, A. and Thomas, J. (2012). An objective, niche-based approach to indicator species selection.*Meth. Ecol. Evo.* 3(2): 317-326.

Case, A. A. (1974). Toxicity of various chemical agents to sheep. *J. Am. Vet. Med. Ass.* 164: 259-277.

Cheeke, P. R. and Shull, L. R. (1985). Natural toxicants in feeds and poisonous plants.AVI Publishing Co., Westport, Connecticut, USA 1-234.

Clarkson, T. W., Zaki, L. A. and Al-Tikriti, S. K. (1976). An outbreak of methylmercury poisoning due to consumption of contaminated grain. *Fed. Proc.* 35 (12): 2395-2399.

Damalas, C. A. and Eleftherohorinos, I. G. (2011). Pesticide exposure, safety issues, and risk assessment indicators. *Intl. J. Environ. Res. & Pub. Heal.*8 (12): 1402-19. DOI:10.3390/ijerph8051402.

Debackere, M. (1983). Environmental pollution: The animal as source, indicator, and transmitter. In: Debackere, M. (Ed.), Veterinary Pharmacology and Toxicology (595-608), AVI Publishing Co., Westport, Connecticut, USA.

Donia, M. D. A., Othman, M. A., Tantawy, G., Khalil, A. Z. and Shawer, M. F. (1974). Neurotoxic effect of leptophos. *Experientia* 30: 63-64.

Galtier, P. and LeBars, J. (1983). Mycotoxin residue problem and human health hazard.In: *Veterin. Pharmacol. Toxicol.* 625-640. AVI Publishing Co.,Westport, Connecticut, USA.

Gupta, S. K., Kincaid, C. T., Mayer, P. R., Newbill, C. A. and Cole, C. R. (1982). A multidimensional finite element code for the analysis of coupled fluid, energy and solute transport. Battelle Pacific Northwest Laboratory PNL-2939, EPA.

Hassan, A., Hamid, F. B. A., Zeid, A. A., Moktar, D. A., Pazek, A. A. and Ibrahain, M. S. (1978). Clinical observations and biochemical studies of humans exposed to leptophos. *Chemosphere* 7: 283-290.

Hill, E. F. and Hoffman, D. J. (1984). Avian models for toxicity testing. *J. Am. Coll. Toxicol.* 3 (6): 357-376.

Hodges, L. (1976). Environmental pollution. 2nd edn, 230-236. Holt, Rinehart andWinston, New York, USA.

Hogan, M., Patmore, L., Latshaw, G. and Seidman, H. (1973). Computer modelng of pesticide transport in soil for five instrumented watersheds. Prepared for the USEnvironmental Protection Agency Southeast Water laboratory, Athens, Ga. by ESL Inc., Sunnyvale, California, USA.

Ingram, D. L. and Mount, L. E. (1975). Man and animals in hot environments. Springer-Verlag, New York, USA.

Irving, L. (1964). Maintenance of warmth in arctic animals. *Symp. Zool. Soc., London*13: 1-34. Cited In: Yousef, M. K., (1985) Stress Physiology in Livestock. vol. II (Chap.10). CRC Press Inc., Boca Raton, Florida, USA.

Jayakumar, R. (2013) Monitoring of Metal Contamination in the Eggs of Two Bird Species in India.*J. Exp Opin. Environ. Biol.* 37 (12): 234-240.

Kati, V., Gasb, T. and Devid, L. (2004). Testing the value of six taxonomic groups as biodiversity indicators at a local scale.*Conserv. Biol.*18(3): 667-675.

Kenaga, E. E. (1978). Test organisms and methods useful for early assessment of acutetoxicity of chemicals. *Environ. Sci. Technol.* 12: 1322-1329.

Lamberth, C., Jeanmart, S., Luksch, T. and Plant, A. (2013). Current challenges and trends in the discovery of agrochemicals. *Science*341 (6147): 742-748. DOI: 10.1126/science.1237227.

Maguire, C. C. and Williams, B. A. (1987). Cold stress and acute organophosphorusexposure: Interaction effects on juvenile northern bobwhite.*Arch. Environ. Contam.Toxicol.* 16: 477-481.

Metcalf, R. L. (1982). Historical perspectiveof organophosphorus ester-induceddelayedneurotoxicity. *Neurotoxicol.* 3 (4): 269-284.

Mikusiński, G., Gromadzki, M. and Chylarecki, P. (2001) Woodpeckers as indicators of forest bird diversity.*Conserv. Biol.*15 (1): 208-217.

Miller, G. T (2004). Sustaining the earth: An integrated approach. Thomson, Brooks Cole Publisher, Pacific Grove, USA 211-216.

National Research Council (NRC) (1972). Lead: Airborne lead in perspective. NationalAcademy of Sciences, Washington, D.C., USA.

National Research Council (NRC) (1979). Interactions of mycotoxins in animal production.National Academy of Sciences, Washington, D.C., USA.

National Research Council (NRC) (1980). Mineral tolerance of domestic animals. NationalAcademy of Sciences, Washington, D.C., USA.

National Research Council (NRC) (1981). Effect of environment on nutrient requirementsof domestic animals. National Academy of Sciences, Washington, D.C., USA.

Neathery, M. W. and Miller, W. J. (1975).Metabolism and toxicity of cadmium, mercuryand lead in animals: a review. *J. Dairy Sci.* 58 (12): 1767-1781.

Nielsen, K. S. (1979). Animal physiology: Adaptation and environment. 2nd Edn. Cambridge University Press, New York, USA.

O'Connell, T. J., Jackson, L. E. and Brooks, R. P. (2000) Bird guilds as indicators of ecological condition in the Central Appalachians.*Ecol. Appl.*10 (6): 1706-1721.

Olawoyin, R., Oyewole, S. A. and Grayson, R. L. (2012). Potential risk effect from elevated levels of soil heavy metals on human health in the Niger Delta. *Ecotoxicol. Environ. Saf.* 85 (1): 120-130.

Ordonez, J. V., Carrillo, J. A., Miranda, C. M. and Yale, J. L. (1966). Organic mercuryidentified as the cause of poisoning in humans and hogs. *Science* 172: 65-67.

Osweiler, G. D., Carson, T. L., Buck, W. B and Van Gelder, G. A. (1985). Clinical and diagnostic veterinary toxicology.3rd Edn. Kendall/Hunt Publishing Co., Dubuque, Iowa, USA.

Picou, J. S. (2009). Disaster recovery as translational applied sociology: Transforming chronic community distress. *Humboldt J. Soc. Rel.* 32(1): 123-157.

Pilastro, A., Awor, A. and Afor, K, (1993) The use of bird feathers for the monitoring of cadmium pollution.*Arch. Environ. Contam. Toxicol.*24(3): 355-358.

Rattner, B. A., Sileo, L. and Scanes, C. G. (1982). Hormonal responses and tolerance to cold of female quail following parathion ingestion. *Pestic. Biochem. Physiol.* 18: 132-138.

Ruckebusch, Y., Toutain, P. L. and Koritz, G. D. (1983). Veterinary pharmacology and toxicology.AVI Publishing Co., Westport, Connecticut, USA.

Selby, L. A., Case, A. A., Dorn, C. R. and Wagstaff, D. J. (1974). Public health hazards associated with arsenic poisoning in cattle. *J. Am. Vet. Med. Ass.*, 165: 1010-1019.

Shull, L. R. and Cheeke, P. R. (1983). Effects of synthetic and natural toxicants on livestock. *J. Anim. Sci.* 57 (2): 330-354.

Smith, J. E. (1982). Mycotoxins and poultry management. *World's Poult. Sci. J.* 38(3): 201-212.

Snyder, C. (2005). The dirty work of promoting recycling of America's sewage sludge. *Intl. J. Occup. Environ. Health.*11 (4): 415-27. DOI: 10.1179/ oeh.2005.11.4.415.

Solomon, F.(2009) Impacts of Copper on Aquatic Ecosystems and Human Health. *Environ. & Commu.* 25-28.

Stolen, E. D., Delton, K. and Wall, O. (2005). Using waterbirds as indicators in estuarine systems: successes and perils. Estuarine Indicators (SA Bortone, Ed.), CRC Press, Boca Raton, Florida, USA 409-422.

Suarez, A. V. and Tsutsui, N. D. (2004). The value of museum collections for research and society.*BioScience* 54 (1): 66-74.

Walker, C. H. (1983). Pesticides and birds-mechanisms of selective toxicity. *Agric. Ecosys. Environ.* 9: 211-226.

Whicker, F. W. and V. Schultz.. 1982. *Radioecology: Nuclear Energy and the Environment.* CRC Press, Boca Raton, Florida, USA.

Yousef, M. K. (1982). Animal production in the Tropics. Praeger Publishers, New York, USA.

Yousef, M. K. (1985a). Stress Physiology in Livestock. vol. I. CRC Press Inc., BocaRaton, Florida, USA.

Yousef, M. K. (1985b). Stress physiology in livestock. vol. II. CRC Press Inc., BocaRaton, Florida, USA.

Assessment of PGPR Bioinoculation in Heavy Metal Contaminated Agriculture Soils and their Impact on Morphological, Physiological, and Yield Behaviors of Crop Plants

Poonam Gusain[1] ,Vir Singh[2], D.P. Uniyal[1] and B.S. Bhandari[3]

[1]Uttarakhand State Council for Science and Technology, Jhajra, Dehradun, 248007, India
[2]Department of Environmental Sciences, G. B. Pant University of Agriculture & Technology, Pantnagar - 263145, India
[3]Department of Botany & Microbiology, HNB Garhwal Central University Srinagar 246174, India

Abstract

This communication attempts to provide a brief review on the impacts of most toxic heavy metals arsenic (As) and cadmium (Cd) and lead (Pb) in different plant species by evaluating changes in morphological physiological and yield characteristics. In addition the severe toxicity symptoms in early growth phase have also been illustrated. Further the utility of plant growth promoting rhizobacteria(PGPR) for enhancing productivity and betterment of soil fertility has been implemented in heavy metal contaminated soils are addressed in this review. However, phytoremediation of heavy metal contaminated soils using hyperaccumulator plant species is still in juvenile phase of the research. This includes the optimization of the process, greater understanding that the plants absorb, translocate, and metabolize heavy metals and plant genetic response to heavy metals and disposal of the metal-laden mass.

Keywords: PGPR, Arsenic, Cadmium, Lead, Phytotoxicity

Heavy metal associated soil pollution

Soil contamination with heavy metals constitutes a major global threat to the environment. Heavy metals often discharged from industrial operations, such as smelting, mining, metal forging, manufacturing of alkaline storage batteries, combustion of fossil fuel, and sewage sludge, and renders contamination of healthy soil environments(Wani *et al.*, 2008). Among heavy metals, which are widespread pollutants to soils are (As), cadmium (Cd) and lead (Pb).

The accumulation of heavymetals into soil leads translocation to plant organs and exerts negative effects on growth and productivity. For instance, the accumulation of metals in plant organs to an undesired level shows limiting effects on physiological processes, inactivating plant protein and consequently adversely affecting the crop yields severely(Wani *et al.*, 2007; 2008).

Arsenic compounds and soil pollution

Arsenic is a ubiquitous element found in earth's crust. Arsenic and inorganic arsenic compounds are well proven toxins to plants, animals and humans and have been used in pesticides, herbicides, preservatives, and pharmaceuticals since long time (Pickering *et al.*, 2000). Calcium and lead arsenates are used as herbicides and insecticides. Application of P fertilizers to soils previously contaminated with lead arsenate has resulted in the release of arsenic to shallow groundwater. Arsenic is also used as a feed additive for poultry as roxarsone for increased growth rates due to its action against intestinal parasites(Mathews, 2011).

Among the toxic metal/metalloid arsenic (As) has received increased attention due to its chronic toxicity, high persistence and increasing input into the environment world over, particularly in developing countries like Bangladesh, China and India (Rai *et al.*, 2011). As has been reported to be strongly phytotoxic and a gamut of toxicity symptoms in plants ranging from reduction in root and shoot biomass, wilting and necrosis of leaf blades, lowered fruit and grain yield andreduction in chlorophyll and photosynthesis. The visual symptoms for arsenic phytotoxicity were observed in mustard plants are characterized by reddish-brown discoloration and necrosis of leaf blades might be due to iron scarcity(Figure.1).

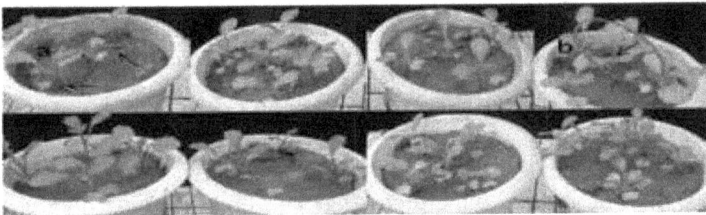

Fig. 1: Effect of heavy metals **a.** inset is showing phytotoxicity in leaves posed by As in mustard plants, **b.** Plants pretreated with metals, recovered after PGPR inoculation (Author's own experimental observations).

Arsenic promotes generation of reactive oxygen species (ROS) which causes oxidative stress and damage to lipid, protein and DNA. For instance: Exposure of *A. annua* to heavy metal have been reported to enhance ROS generation which augments artemisinin yield by facilitating rapid conversion of artemisinic acid to artemisinin(Rai *et al.* 2011).

Plant Growth Promoting Rhizobacteria (PGPR)

The use of beneficial microbes in agricultural systems to improve plant and soil health is increasing evidence that beneficial microbes can enhance plants' tolerance to adverse environmental stresses such as heavy metal contaminations (Sheng, 2005). PGPR, generally refers to a group of soil and rhizosphere free-living bacteria colonizing roots in a competitive environment and exerting a beneficial effect on plant growth (Kloepper *et al.*, 1989; Bakker *et al.*, 2007).Among PGPRs representatives *Acinetobacter, Agrobacterium, Arthrobacter, Azoarcus, Azospirillum, Azotobacter, Bacillus, Burkholderia, Enterobacter, Klebsiella, Pseudomonas, Rhizobium, Serratia,* and *Thiobacillus* are widely accepted genera. However, *Pseudomonas* and *Bacillus* species together with *Streptomyces* species, constitute the most important group of bacteria often found in the rhizosphere of many crop plants(Adesemoye and Egamberdieva 2013).

The effect of PGPR inoculants on crops

Lee *et al.* (1986) observed the symptoms of As toxicity in rice plants. The biomass production and yields of a variety of barley (*Hordeum vulgare* L.) and ryegrass (*Lolium perenne* L.) are reduced significantly at 50mg As kg^{-1} arsenic concentrations to soil (Jiang and Singh, 1994). Duman *et al.* (2004) revealed that at a high concentration (higher than 800µg l^{-1}), the growth rate of *L. gibba* was significantly inhibited after exposure to As(III).Report by Seth *et al.* (2007), demonstrated that arsenic species can have a hormesis effect on photosynthetic pigments. A hormesis is a stimulatory effect that is induced by low doses of an otherwise toxic substance (Duman *et al.*, 2010).

Because of the chemical similarity between arsenate and phosphate, the P/As ratio in plants is important in regulating plant arsenate uptake and toxicity. Meharg and Macnair (1991)observed an As uptake reduction of 75% at 0.5 mMphosphate in both tolerant and nontolerant plant genotypes of grass (*Holcus lanatus*). Arsenate concentrations in alfalfa (*Medicago sativa* L.) shoots are also reduced by phosphate (Khattak *et al.*, 1991). For Indian mustard (*Brassica juncea* L.), grown in 0.5 mMarsenate and 1 mMphosphate, 55–72% reduction of arsenate uptake over the control has been reported (Meharg and Macnair,1992). At high levels of arsenate (30 mM/kg), however, phosphate does not overcome arsenate toxicity even at a molar P/As ratio of 24. This suppression is because of a feedback regulation of the arsenate/phosphate transporter, i.e. reduced arsenate uptake through the suppression of the high-affinity uptake system (Mcharg and MacNair, 1992). Winter wheat cultivar Jing 411 had higher shoot biomass, but lower percentages of biomass allocated to roots than cultivar Lovrin 10 (P < 0.001) when exposed to 667 lM arsenic (Geng *et al.*, 2005).

Cadmium and phytotoxicity effects

Cd is easily translocated from plant roots to above ground tissues, and poses a serious health issue to humans through the food chain. Cd toxicity causes oxidative stress, which can take place possibly by generating reactive oxygen species (ROS) such as superoxide radicals (O_2-), singlet oxygen ($^1O_2$), hydrogen peroxide (H_2O_2) and hydroxyl radicals(OH−)(Zhao 2011; Millan *et al.* 2009).Cd in plants interferes with physiological processes viz, decrease carbon assimilation, induces stomatal closure and disturb plant water status, inhibits chlorophyll synthesis, damages root tips, reduce nutrient uptake, impairs photosynthesis,inhibits plant growth and generates oxidative stress(Zhao 2011).

Effect of PGPR inoculants on Cd treated plants

The effects of Cd have been investigated in tomato (*Lycopersicon esculentum*) plants grown in a controlled environment in hydroponics, using Cd concentrations of 10 and 100µM (Millan *et al.* 2008). Plant growth was reduced in both Cd treatments, leaves showedchlorosis symptoms when grown at 10µM Cd and necrotic spots when grown at 100µM Cd, and root browning was observed in both treatments (Millan *et al.* 2008).Cadmium has been shown to be one of the most effective inhibitors of photosynthetic activity. It can enter chloroplasts and disturb chloroplast function by inhibiting the enzymatic activities of chlorophyll biosynthesis, pigment–protein complexes, O_2-evolving reactions of photosystem II, electron flow around photosystem I and chloroplast structure (Shukla *et al.*, 2008).

Tripathi *et al.* (2005) reported 39.14% reduction in chlorophyll content was regained by siderophore producing *Pseudomonas putida* KNP9 in the presence of Cd in mung bean. Sinha and Mukharjee (2008), also demonstrated that *Pseudomonas aeruginosa* strain KUCd1 reduced 59.22% of Cd uptake in roots and 47.40% of Cd in the shoots of pumpkin plants. Their findings were also significant in the case of mustard plants, where the strain was also found to reduce 52.44% and 36.89% of Cd uptake in both roots and shoots, respectively.

Another study indicated that the structure of cell membrane is affected seriously and the mitochondrion crista disappeared with the Cd stress (Li, 2000).

Inoculation of canola (*Brassica napus*) with metal-resistant PGPR containing ACC deaminase stimulated growth of plants cultivated in Cd contaminated soil (Belimov *et al.*, 2001). In addition, various N_2-fixing and auxin-producing PGPR immobilized Cd and promoted growth and nutrient uptake by barley plants in the presence of toxic Cd concentrations (Belimov and Dietz 2000; Pishchik *et al.*, 2002). Belimov *et al.* (2005) isolated PGPR strains included *Variovorax paradoxus*, *Rhodococcus* sp. and *Flavobacterium* sp., from the root zone of Indian mustard grown in Cd-supplemented soils as well as sewage sludge and mining waste highly contaminated with Cd. The isolated strains were capable of stimulating root elongation of *B. juncea* seedlings either in the presence or in the absence of toxic Cdconcentrations. Several Cd-tolerant PGPR strains containing ACC deaminase have been also isolated from the rhizoplane of Indian mustard and pea seedlings (Belimov *et al.*, 2001).

Mia *et al.* (2012) suggested that PGP inoculation enhanced the seed emergence and contribute to the protection of plants against certain pathogens and pests due to siderophore-mediated antibiosis where germination increased slightly while the seeds were inoculated with PGPR (Figure 2.).

Fig. 2: Effect of Cd on seed germination A. Seed germination in control plants B. Seed germination in plants pretreated with PGP

Lead toxicity

Among common pollutants that adversely affect plants, lead is one of the most toxic and frequently encountered, used in batteries, ceramics, chemicals and fertilizers, gasoline, hair dyes, glass, newsprint, paints, pesticides, pottery and rubber toys. Lead can persist in the environment for 150-5000 years. Once in water it enters the food chain and adversely affects the flora and fauna. In plants, enhanced level of lead in soil caused significant reduction in plant height, root-shoot ratio, dry weight, nodule per plant, chlorophyll content in *Vigna radiata* (Tomar *et al.*, 2000) and in *Menthaspicata* (Bekiaroglou and Karatagli, 2002). Lead toxicity may also cause swollen, bent, short and stubby roots that show an increased number of secondary roots per unit root length (Pourrut *et al.*, 2011).

Jiang and Liu (2010) reported mitochondrial swelling, loss of cristae, vacuolization of endoplasmic reticulum and dictyosomes, injured plasma membrane and deep colored nuclei, after 48–72 h of lead exposure to *Avena sativum* roots. In addition Pb is known to reduceseed germination, seedling development,transpiration, chlorophyll production, lamellar organization in the chloroplast,and cell division in plants grown in Pb affected soils (Gupta *et al.*, 2010).

Lead adsorption into roots has been documented in several plant species including*Vigna unguiculata, Festuca rubra, Brassica juncea, Lactuca sativa*and *Funaria hygrometrica.*For most plant species, the majority of absorbed lead (approximately 95% or more) is accumulated in the roots, and only a smallfraction is translocated to aerial plant parts, as has been reported in *Vicia faba, Pisumsativum, Phaseolus vulgaris, Vigna unguiculata, Nicotiana tabacum,Lathyrus sativus, Zea mays, Avicennia marina* (Pourrut *et al.*, 2011).

Lead-induced inhibition of seed germination has been reported in *Hordeum vulgare, Elsholtzia argyi, Spartina alterniflora, Pinus halepensis, Oryza sativa* and *Zea mays.* Inhibition of germination may result from the interference of lead with protease and amylase enzymes (Sengar *et al.*, 2009). Lead exposure in plants also strongly limits the development and sprouting of seedlings (Pourrut *et al.*, 2011).

Mesmar and Jaber (1991) reported a strong reduction in seed germination of wheat and lentil by about 60% with higher concentration of lead (20mM). Bose and Bhattacharya (2008) have distinguished the total metal concentrations in 7-day-old wheat plant seedlings. The total metal concentration in shoot of wheat seedling followed the trend Fe > Zn >Mn > Pb > Ni > Cr > Cu > Cd, when the wheat plants were grown in sludge-amended soils. In addition Tripathi *et al.* (2005), reported 18.7%, 19.8%,48%, 35% decline in shoot length, root length, wetweight, and dry weight respectively in mung bean plants due to lead toxicity. Pb inhibited the respiration strength remarkably during seed germination in rice in a concentration dependent manner (Li, 2000).

A high proportion of metal resistant bacteria persist in the rhizosphere of the hyperaccumulators *Thalaspi caerulescens* and *Alyssum bertolonii*grown in soil contaminated with Zn and Ni, respectively. Inoculation of Indian mustard and canola (*Brassica campestris*) seeds with the plant growth-promoting rhizobacteria (PGPR) strain *Kluyvera ascorbata* SUD165, which produces siderophores and contains the enzyme 1-aminocyclopropane-1-carboxylate (ACC) deaminase, protectedthe plants against Ni, Pb and Zn toxicity (Belimov *et al.,* 2005). Seed germination and respiratory rate in soybean are also observed to be repressed by the pollution of Cu and As (Li, 2000). Indeed, there is a large variation in the sensitivity of plant species to trace metals even among populations within a species. Reichman *et al.* (2004) reported that shoot mass was reduced by 10% at 5.0 lM Mn for *Eucalyptus crebra* but only at 330 lM Mn for *Eucalyptus camaldulensis* Dehnh. Similarly, Edwards and Asher (1982) found that across 13 crop and pasture species, the external Mn concentration needed to reduce plant dry mass by 10% varied from 1.4 lM in two monocots (maize and wheat) to 65 lM in a dicot (sunflower). In a study with *Silene cucubalus*, de Vos et al. (1991) reported that the EC_{50} for root elongation was 4.0 lM Cu in a sensitive population but was 150 lM Cu in a tolerant population collected from a Cu-contaminated site.

Conclusion

Heavy metals have gained considerable attention as persistent toxic pollutants of concern,largely because of the growing anthropogenic pressure on the environment. PGPR approachhas emerged as a cost-effective and environment-friendly technology in cleaning-up contaminated soils and promoting plant health and vigour growth. More research efforts are needed to make PGPR approach a practical technology to clean up metal contaminated soils.

Acknowledgment

One of the authors (P.G) is thankful to the Department of Science and Technology for the INSPIRE fellowship to work at the GB Pant University of Agriculture and Technology Pantnagar India.

References

Adesemoye, A.O. and Egamberdieva, D. (2013). Beneficial effects of plant growth-promoting rhizobacteria on improved crop production: prospects for developing economies. In D.K. Maheshwari (Ed.), Bacteria in Agrobiology: Crop Productivity(pp-). Heidelberg: Springer-Verlag Berlin.

Bakker, P.A.H.M., Raaijmakers, Bloemberg, Hofte, Lemanceau, and Cooke, M. (2007). New perspectives and approaches in plant growth-promoting rhizobacteria research. *Eur J Plant Pathol.* 119:241–242.

Belimov, A.A., Dietz, K.J. (2000). Effect of associative bacteria on element composition of barley seedlings grown in solution culture at toxic cadmium concentrations. *Microbiol. Research.* 155: 113–121.

Belimov, A.A., Safronova, Sergeyeva, Egorova, Matveyeva, Tsyganov, Borisov, Tikhonovich, Kluge, Preisfeld, Dietz and Stepanok, V.V. (2001). Characterisation of plant growth-promoting rhizobacteria isolated from polluted soils and containing 1-aminocyclopropane-1-carboxylate deaminase. *Can J of Microbiol.* 47: 642–652.

Belimov, A.A., Hontzeas, Safronova, Demchinskaya, Piluzza, Bullitta and Glick, B. R. (2005). Cadmium-tolerant plant growth-promoting bacteria associated with the roots of Indian mustard (*Brassica junceaL.* Czern.). *Soil BiologBiochem.* 37:241–250.

Bekiaroglou, P., and Karatagli, S. (2002). Effect of lead and zinc on *Mentha spicata*. J Agron and Crop Sci. 188(3): 201-205.

de Vos, C.H.R., Schat, Dewaal, Vooijs, and Ernst W.H.O. (1991). Increased resistance to copper-induced damage of the root cell plasmalemma in copper tolerant *Silene cucubalus. Physiol Plantarum.* 82:523–528.

Duman, F., Ozturk and Aydin, Z. (2010). Biological responses of duckweed (*Lemna minor* L.) exposed to the inorganic arsenic species As(III) and As(V): effects of concentration and duration of exposure. *Ecotoxicology.* 19:983–993.

Geng, C-N, Zhu, Tonga, Smith and Smith F.A. (2005). Arsenate (As) uptake by and distribution in two cultivars of winter wheat (*Triticum aestivum* L.). *Chemosphere.* 62:608–615.

Gupta, D., Huang, Yang, Razafindrabe, and Inouhe, M. (2010) The detoxification of lead in *Sedum alfredii* H. is not related to phytochelatins but the glutathione. *J Hazard Mater.* 177(1–3):437–444.

Jiang, Q.Q., and Singh, B.R. (1994). Effect of different forms and sources of arsenic on crop yield and arsenic concentration. *Water Air Soil Pollut.* 74:321–343.

Jiang, W., and Liu, D. (2010). Pb-induced cellular defense system in the root meristematic cells of *Allium sativum* L. *BMC Plant Biol.* 10:40–40.

Khattak, R.A., Page, Parker and Bakhtar, D. (1991). Accumulation and interactions of arsenic, selenium, molybdenum and phosphorus in alfalfa. *J. Environ. Qual.* 20: 165–168.

Kloepper, J.W., Lifshitz and Zablotwicz, R.M. (1989). Free-living bacterial inocula for enhancing crop productivity. *Trend Biotechnol.* 7:39–43.

Lee, M. H., Lim and Kim, B.K. (1986). Behavior of Arsenic in paddy soils and effects of absorbed arsenic on physiological and ecological characteristic of rice plant. II. Effect of As treatment on the growth and as uptake of rice plant. *Korean J. Environ. Agric.* 5:95-100.

López-Millán, A., Sagardoy, Solanas, Abadía and Abadía, A. (2009). Cadmium toxicity in tomato (*Lycopersicon esculentum*) plants grown in Hydroponics. *Environmental and Experimental Botany.* 65:376–385.

Mathews, S. (2011). (Ph d thesis) Arsenic hyperaccumulation by *Pteris vittata* l.-arsenic transformation, uptake and environmental impact. University of Florida.

Meharg, A. A. and MacNair, M. R. (1991). Uptake, accumulation and translocation of arsenate in arsenate-tolerant and non-tolerant *Holcus lanatus L. New Phytol.* 117:225–231.

Meharg, A.A. and MacNair, M.R. (1992). Suppression of the high affinity phosphate uptake system: a mechanism of arsenate tolerance in *Holcus lanatus L. J.Exp. Bot.* 43:519–524.

Mesmar, M.N., Jaber, K. (1991). The toxic effect of lead on seed germination, growth, chlorophyll and protein contents of wheat and lens. *Acta Biol Hung.* 42:331–344.

Mia, M.A.B., Shamsuddin and Mahmood, M. (2012). Effects of rhizobia and plant growth promoting bacteria inoculation on germination and seedling vigor of lowland rice. *African J Biotechnol.* 11:3758–3765.

Pickering, I.J., Prince, George, Smith, George and Salt, D.E. (2000). Reduction and coordination of arsenic in Indian mustard. *Plant Physiol.* 122:1171–1177.

Pishchik, V.N., Vorobyev, Chernyaeva, Timofeeva, Kozhemyakov, Alexeev, and Lukin, S.M. (2002). Experimental and mathematical simulation of plant growth promoting rhizobacteria and plant interaction under cadmium stress. *Plant and Soil.* 243:173–186.

Pourrut, B., Muhammad, Dumat, Winterton and Eric Pinelli. (2011). Lead Uptake, Toxicity, and Detoxification in Plants. *Rev Environ Contam Toxicol.* Springer Verlag, 213:113-136.

Reichman, S.M., Menzies, Asher and Mulligan, D. (2004). Seedling responses of four Australian tree species to toxic concentrations of manganese in solution culture. *Plant and Soil.* 258: 341–350.

Rai, R., Pandey and Rai, S. (2011). Arsenic-induced changes in morphological, physiological, and biochemical attributes and artemisinin biosynthesis in *Artemisia annua*, an antimalarial plant.*Ecotoxicology.* 20:1900–1913.

Sengar, R.S., Gautam, Sengar, Sengar, Garg, Sengar, and Chaudhary, R. (2009). Lead stress effects on physiobiochemical activities of higher plants. *Rev Environ Contam Toxicol.*196:1–21.

Seth C.S., Chaturvedi and Misra, V. (2007). Toxic effect of arsenate and cadmium alone and in combination on giant duckweed (*Spirodelapolyrrhiza*L.) in response to its accumulation. *Environ Toxicol.* 22:539–549.

Shao, Y., Jiang, Zhang, Ma and Li, C. 2011. Effects of arsenic, cadmium and lead on growth and respiratory enzymes activity in wheat seedlings.*Afr J of Agric Res*. 6(19): 4505-4512.

Sheng, X.F. (2005) Growth promotion and increased potassium uptake of cotton and rape by a potassium releasing strain of *Bacillus edaphicus*. *Soil Biol Biochem*. 37:1918–1922.

Shukla, U.C., Murthy and Kakkar, P. (2008) Combined effect of ultraviolet-B radiation and cadmium contamination on nutrient uptake and photosynthetic pigments in *Brassicacampestris* L. seedlings. *Environ Toxicol*. 23:712–719.

Sinha, S., Mukherjee, S.K. (2008). Cadmium induced siderophore production by high Cd resistantbacterial strain relieved Cd toxicity in plants through root colonization. *Curr Microbio*. 56:55–60.

Tomar, M., Kaur, Neelu and Bhatnagar A.K. (2000). Effect of enhanced lead in soil on growth and development of *Vigna radiata* (L.) Wilezek. Indian J. Plant Physio. 5(1): 13-18.

Tripathi, M., Munot, Shouche, Meyer and Goel R. (2005). Isolation and functional characterization of siderophore producing lead and cadmium resistant *Pseudomonas putida* KNP9. *Curr Microbiol*.50:233–237.

Wani, P.A., Khan and Zaidi, A. (2007). Cadmium, chromium and copper in green gram plants. *Agron Sustain Dev*. 27:145–153.

Wani, P.A., Khan and Zaidi, A. (2008). Effect of metal-tolerant plant growth-promoting rhizobium on the performance of pea grown in metal-amended soil.*Arch Environ Contam Toxicol*. 55:33–42.

Zhao, Y. (2011). Cadmium accumulation and antioxidative defenses in leaves of *Triticum aestivum* L. and *Zea mays* L.*Afr J Biotechnolog*. 10(15): 2936-2943.

A Review on Contamination of Soil through Heavy Metals

Veethika Tilwankar, Swapnil Rai and S. P. Bajpai

Department of Environmental Science, Amity University Madhya Pradesh, Gwalior, 474005 – India

Abstract

Due to mismanaged Industrial activities the concentrations of different heavy metals in environment have increased which are now disturbing the biogeochemical cycles. Soil is the basic and most essential component of our environment and act both as the source and the sink for different pollutants. Heavy metals inhibit microbial or chemical degradation, metals can only change their chemical forms but does not undergoes biodegradation. Excessive use of heavy metals led to the contamination of soil which is toxic to plants, animals, and humans. Metals once enter in the environment accumulate in soil, water and atmosphere and causes deteriorating and lethal effect. To protect our environment proper remediation and their speciation, bioavailability, characterization is necessary. By knowing the speciation of heavy metals good remediation techniques should be chosen for its proper elimination from environment. Due to excess use these days heavy metals are been studied widely for its toxic effects. Exposure to heavy metals like mercury (Hg), lead (Pb), cadmium (Cd), copper (Cu), arsenic (As), chromium (Cr) for a long time can cause disastrous effects on human, animals and plants. This paper reviews the accumulation of these metals in soil and its interaction and toxic effects on human, plant and animal life.

Keywords: Bioaccumulation, Heavy Metals, Toxic, Contamination

Introduction

Heavy metals are the metals found in earth crust naturally and are heavier than water. In developing countries like India due to indiscriminate anthropogenic activities there is increased heavy metals pollution these days. Heavy metals are essential in small quantity for humans, for plant growth, but are proved to be toxic when present in higher quantity. Some heavy metals are essential while some are non essential, some common heavy metals that cause toxicity are lead (Pb), cadmium (Cd), mercury (Hg), arsenic (As), copper (Cu), chromium (Cr). Heavy metal pollution has increased at a larger extent and results in the deterioration of environmental quality and human health (Han *et al.*, 2002; Zojaji *et al.*, 2014; Sayadi and Rezai *et al.*, 2014). Heavy metals also sometimes proved to be toxic even when present in small quantity. The property of heavy metals is to accumulate in the system and then enters in to food chain. Once it entered in the soil, water and vegetable it cannot be easily removed as these are non-biodegradable. Metal toxicity caused by any reason it may be natural or anthropogenic and it affects ecology, evolution, nutrition and overall environment balance (Nagajyoti *et al.,* 2010; Jaishankar *et al.,* 2014).

In past, due to inadequate knowledge, inadequate handling of chemicals and with no regulations, led to many such disastrous incidents. The common among them are Itai-Itai and Minimata. These incidents are raised awareness among the people about the toxicity of heavy metals. Dietary substances are contaminated by chemicals and some non-essential heavy metals and have many toxic effects on the health of humans and animals (D'souza and Peretiatko, 2002). Heavy metals enter into the body through various routes like inhalation, ingestion etc. After entering in the body they causes malfunctioning of various organs, it displaces the necessary metal from their place and inhibit their activities. They bind with proteins and DNA and thus disturbs functioning of body organs (Flora *et al.,* 2008). All the heavy metals have common organs to affect as they attacks CNS (central nervous system), kidney, liver (Flora *et al.,* 2008; Jan *et al.,* 20011). Heavy metals are widely been used in industries, agriculture and domestic activities and results in toxic health effects on humans, animals and also destroy plants growth and disturbs the ecological balance. It affects multiple organ systems. Toxicity of heavy metals depends on exposure time, route of entry and susceptibility of particular human (Jan *et al.,* 2011). In India, many cities generates industrial waste and dumped in soil. Govil et al 2001 reported soil contaminated with V, Fe, As, Cd ,Ba, Se, Cr, Zn, Sr, Cu were found very high in and around patancheru industrial development area of Andhra Pradesh of Hyderabad city. Krishna and Govil (2004) reported high level of heavy metals like Cr, Cu, Pb, Sn,V, Zn in soils collected from pali industrial area of Rajasthan.

Heavy metals in environment

Though heavy metal is present in the soil due to weathering as trace metals and are rarely toxic (Piezzynski *et al.,* 2000; Jan *et al.,* 2011) Increased human activities in the environment led to the accumulation of heavy metals and thus it's increased concentration to a limit that would be hazardous to plants, animals and human beings (D'Amore *et al.,* 2005). Heavy metals become pollutant when their concentration rises in the environment. Heavy metals in the environment is the sum total of pollutants comes from parent rock, atmospheric fallout, fertilizers, agricultural source, inorganic contaminants and organic matter (Smith *et al.,* 1995)

$$M_{(total)} = (M_p + M_a + M_f + M_{ag} + M_{ow} + M_{ip}) - (M_{cr} + M_l)$$

Where, M=heavy metal, p=parent rock, a=atmospheric fallout =fertilizers, ag=agricultural sources, ow=organic matter, ip=inorganic pollutant, cr=crop removal, l =leaching. Heavy metals are bio available and more mobile when present in environment from anthropogenic sources (Kuo *et al.*,1983; Kaasalainen and Yli-Halla, 2003). It comes in environment from various sources such as disposed untreated industrial waste ,leaded gasoline ,paints, fertilizer, pesticides, coal combustion petrochemicals etc. some of the sources are discussed here:

Fertilizers: - To grow more crops and to fulfill need for a large population fertilizers are used. For a better crop soils must be provided with the essential macro nutrient (N, P, K, Ca) and some micro nutrient (Co, Cu, Fe ,Mn ,Mo, Ni, Zn)which are essential for the growth of good crops(Lasat, 2000).For this reason soils are supplied by these nutrient through the fertilizers which contain these micro nutrient. Thus by continuously adding fertilizers though in small quantity but at regular interval leads to some deposition of heavy metals in the soil as well and thus increases its concentration.

Pesticides:- Pesticides generally contains some heavy metals and are used in fields for removing pest and insects. For example Bordeaux mixture which is combination of copper sulphate and copper oxychloride used as pesticides (Jones and Jarvis, 1981) .To control parasites and insect's lead arsenate was used, with extensive use of pesticides the background concentration of heavy metals exceeds.

Wastewater :- Due to water scarcity in many countries now a day's wastewater which comes from municipal and industries is used to irrigate the agricultural fields (Reed *et al.*, 1995). Though concentration of heavy metals in the treated industrial water is less but due to long term irrigation with wastewater leads to accumulation of heavy metals in soil.

Metal mining:- In the mining industries ,mine tailings settles at the bottom after mining and directly discharged in the deep soil and underground(Devolder *et al.*, 2003). Zinc and lead are the main heavy metals found in the mine discharge which accumulate in soil and have significant health risk to humans. Other industries like textile, petrochemicals, tanning also discharge their waste in soil and thus by small concentration the deposition of heavy metals exceeds its background limit and soil becomes contaminated.

Atmospheric fallouts:-Airborne sources include the particulate comes in the form of dust, vapour, gas, vehicular emission and also emission from stacks. Emissions are of two kinds one is stack emission which covers a larger area and the other one is fugitive emission which covers smaller area. All the particulates and pollutants from fires, through thermal plants, factory chimney generally got deposited on land, plants, rivers and ponds. Type of heavy metal and its concentration both depends on the source from which it emitted .For example, Zn, Pb, Cd found in plants where there was a smelting plant. Combustion of petrol is the source of lead in atmosphere(USEPA, 1996).

Heavy metals and their related risks:

Commonly Pb, Hg, Cu, As, Cr, Cd are found in abundance in contaminated soils. For understanding the basic chemistry, its interaction in soils, bioaccumulation, and biomagnifications of heavy metals, its properties and risk should be discussed. Once the soil got contaminated with heavy metals they got absorbed then they distributed in different chemical forms and become mobile in food chain. Due to their toxicity and risk to humans, plants and animals heavy metals are studied worldwide (Shiowatana *et al.*,2001; Buekers, 2007)

Lead(Pb):-Lead lies in group IV and sixth period of the periodic table having atomic number 82,density 11.4gcm^{-3} ,atomic mass 207.2,melting point 327.4°c and boiling point 1725°c. It is bluish grey in colour and found with sulphur and oxygen [PbS,PbSO$_4$] and [PbCO$_3$]. In earth crust it ranges from 10 to 30 mgkg^{-1}(USDHHS, 1999).Lead in soil average ranges about 32 mg kg^{-1} worldwide i.e. it ranges from 10 to 67 mg kg^{-10}(GWRTAC, 1997). Lead (Pb) commonly used in manufacturing of batteries, plumbing, bearings, solders etc. Generally lead which is found in soil, water and groundwater are the complex form of lead such as lead oxide and hydroxide. Reactive form of lead is lead (II) (NSC, 2009). Lead (II) are ionic in nature eg Pb^{2+}. Some salts forms of lead such as Pb(OH)$_2$, 2PbCO$_3$ which is used in paint and become a source of lead poisoning in small children who ate paint. Ingestion and inhalation is the common route of exposure to lead compounds. Lead (Pb) affects brain and CNS which leads to plumbism or death. Kidney and gastrointestinal tract also affected by lead. Small children are at risk because lead accumulation their development got impaired and suffers from lower IQ, and also sometimes shows hyperactivity (Smith *et al.*, 1995).Lead is one of the most toxic metals and is extensively studied for its lethal effects. Accumulation and poisoning of lead depends on the level of lead been ingested and secondly the duration for which it has been taken. Concentration of lead in leafy vegetables and rooted crops are found to be higher. When the concentration of lead increases the risk of contamination and its distribution through food chain also increases.

Chromium:- Chromium lies in group VI B of the periodic table having atomic number 24,density -7.19g cm^{-3}, atomic mass 52,melting point 1875°c and boiling point 2665°c.It is not found naturally. Chromium is found as chromites FeCrO$_4$ .It is released from electroplating industries (Scragg, 2006).Chromium (IV) is dominant in contaminated soil and it also include chromate (CrO^{2-}) and dichromate (Cr$_2$O$_7^{2-}$),it gets adsorb on soil surfaces especially,iron and aluminium oxides. Chromium (VI) is toxic and has greater mobility. Mobility of chromium depends on sorption properties of soil, iron oxide content, and the amount of organic matter present in soil. Some of the chromium which is unabsorbed leached directly in the groundwater from soil. When the soil pH increases the leach ability of chromium (IV) increases. Allergic dermitis in humans is caused by chromium.

Arsenic:- Arsenic lies in group VA and period 4 of periodic table. It is present as the residue of coal combustion .Arsenic have atomic number 33, density 5.72gcm^{-3}, melting point 817°c and boiling point 613°c. It is found as arsenate As(V).It has a complex chemistry occurs in several oxidation state (III,0,V) (Scragg, 2006). In extreme reducing situation elemental arsenic and arsenic AsH$_3$ is present. Arsinite adsorb with sulphides and other sulphur compounds. It forms methylarsinic acid $(CH_3)_2AsO_2H$ as arsenic speciation. Arsenic causes skin damage and central nervous system damage and are probable carcinogen (Campbel, 2006)

Cadmium:- Cadmium lies at second row of transition metals having atomic number 48, density $8.65gcm^{-3}$, atomic mass 112.4, melting point 320.9°c, boiling point 765°c.It lies below zinc in the periodic table and having similar properties as Zn. Zinc is known as necessary micro nutrient for plants and animals for growth. Sometimes zinc being displaced by cadmium and it may cause disturbance in metabolic activities (Manahan, 2003). Cadmium commonly found used in Ni/Cd batteries as secondary power sources, used for coating vessels and vehicle part as anti corrosive .It is used as stabilizers and pigments in electronic compounds. It is found in many products as impurities such as petroleum products, detergents etc. Cd is found as a by product of Zn and lead refinery. By the industrial disposal, atmospheric deposition, agricultural runoff the concentration of Cd increases in soil. Cadmium have some toxicology properties when absorbed by any organism and plants it remains there for longer duration and become accumulated and persistent in the food chain proteinuria a renal disease which is caused by cadmium (Cd) affect enzymes activity and also activity of acid synthetas, oxysulfatase, alcohol dehydrogenase (Sumner, 2000) One incident of cadmium poisoning was reported in Japan near fuchu in the Jintsu river valley by dietary ingestion of cadmium. The disease known as itai-itai which also means ouch-ouch in Japanese. Kidney abnormalities and osteomalacia (bone disease) are the result of itai-itai. Cadmium is also inhaled through cigarette and tobacco smoking by which it enters in the body (Sumner, 2000)

Copper :- Copper lies in group IB and in period 4 of periodic table having atomic number 29,density $8.96gcm^{-3}$,atomic mass 63.5,melting point 1083°c,boiling point 2595°c.Copper is essential micronutrient for the growth of plant and animals and is also help in making blood haemoglobin in humans .It is required for disease resistant and seed production in plants. If present in high quantity copper causes anaemia, kidney and liver damage. Contamination of soil by these trace metals have both direct and indirect effects, directly it effects the growth of plants and yield of crops indirectly it enters in the food chain.

Mercury :- Mercury lies in same group as cadmium .It is liquid metal with atomic number 80, density $13.6g\ cm^{-3}$, atomic mass 200.6, melting point 13.6°c, boiling point 357°c. Mercury is released from coal combustion. It also released from manometers, gas/oil pipelines and also from pressure measuring stations. Mercury exist as mercurous (Hg^{2+}), mercuric (Hg^{2+}), elemental (Hg^0) in environment. When reducing condition exists organic mercury reduced to elemental form and then in alkylated forms via biotic and abiotic means. Mercury is highly volatile in air and highly toxic in nature (Scragg, 2006).It forms strong complexes with organic, inorganic compounds of aquatic system and becomes more soluble. Mercury affects kidney when its concentration exceeds in human body.

Soil metal concentration standards

Concentration of metals in soil depends on the site, activities at the site, disposal of waste at site, form of waste, contaminants concentration and their interaction with soil (NSC, 2009). Metals present in soil may be of solid ,liquid and gaseous state and determination of this metal has been done by elemental analysis and expressed in mgkg^{-1} metal soil. To find out the leach ability of metals and level of contamination one more test is used called as toxicity characteristic leaching procedure(TCLP) (USEPA method 1311) .Concentration of pollutants measured as total dissolved metals in mgl^{-1}or gl^{-1} or in molar concentration (mol L^{-1}). According to some soil standards followed in India as by CPCB, these are the threshold limits of soil heavy metals and above which the concentration of heavy metals proved toxic and adversely affecting the environment. Table 1 shows safe limit of heavy metals in soil.

Table 1: SAFE LIMIT OF HEAVY METALS IN SOIL (mgkg^{-1})

Standards	Fe	Zn	Cu	Pb	Cd	Mn	Cr
INDIAN STANDARDS (AWASTHI 2000)	NA	300-600	135-270	250-500	3-6	NA	NA
EUROPEAN UNION STANDARDS EU)2002	NA	300	140	300	3.0	NA	150
USEPA 2010	NL	200	50	300	3.0	80	NA
KABATA-PENDIAS 2010	1000	NA	NA	NA	NA	NA	NA

Source:CPCB

Conclusion

Heavy metals are the toxic which is ruining our environment and have lethal and disastrous effects on human, plant and animal. For the remediation of heavy metal contamination in soil, water and vegetables its basic properties, chemistry and probable sources must be understood for applying suitable technique to remediate this contamination. This is very important and essential to remove this contaminant from environment in order to reduce its lethal effects to maintain food quality. Heavy metals affect our ecosystem if present in huge quantity and soil, water act as the primary source for contamination and thus it slowly spreads in the environment through food chain thus proper remediation techniques should be used to stop its lethal and toxic effects. Basic knowledge of heavy metals, its properties and its risk are discussed in this paper.

References

Buekers, J. (2007). Fixation of cadmium, copper, nickel and zinc in soil: kinetics, mechanisms and its effect on metal bioavailability,Ph.D. thesis, Katholieke Universiteit Lueven, Dissertationes De Agricultura, Doctoraatsprooefschrift nr.

Campbell, P.G.C.(2006). Cadmium-A priority pollutant. *Environmental Chemistry*, 3(6): 387–388.

D'Amore ,J. J., Al-Abed ,S.R, Scheckel, K. G., and Ryan J. A.(2005). Methods for speciation of metals in soils: a review. *Journal of Environmental Quality*. 34 (5): 1707–1745.

D'Souza, C. and Peretiatko, R. (2002). The nexus between industrialization and environment: A case study of Indian enterprises. *Environ. Manag. Health.*, 13: 80–97.

DeVolder, P. S., Brown, S. L., Hesterberg, D. and Pandya, K.(2003). Metal bioavailability and speciation in a wetland tailings repository amended with biosolids compost, wood ash, and sulfate. *Journal of Environmental Quality*, 32 (3): 851–864.

Flora, S.J.S., Mittal, M. and Mehta, A.(2008). Heavy metal induced oxidative stress and its reversal by chelation therapy. *Ind. J. Med. Res.*, 128: 501–523.

Govil, P.K., Reddy, G.L.N. and Krishna, A.K. (2001). Contamination of soil due to heavy metals in the Patancheru industrial development area, Andhra Pradesh, India. *Environmental Geology*. 41: 461-469.

GWRTAC.(1997). Remediation of metals-contaminated soils and groundwater, Tech. Rep. TE-97-01, GWRTAC, Pittsburgh, Pa, USA, GWRTAC-E Series

Han, F.X., Banin, A, et al.(2002). Industrial age anthropogenic inputs of heavy metals into the pedosphere. Naturwissenschaften, 89(11): 497-504

Jaishankar, M., Mathew, B.B., Shah, M.S., Murthy, K.T.P. and Gowda, S.K.R. (2014) Biosorption of few heavy metal ions using agricultural wastes. *J. Environ. Pollut. Hum. Health.* 2: 1–6.

Jan, A.T.,Ali, A. and Haq, Q.M.R.(2011). Glutathione as an antioxidant in inorganic mercury induced nephrotoxicity. *J. Postgrad. Med.*, 57:72–77.

Jones, L. H. P. and Jarvis, S. C.(1981). *The fate of heavy metals* in The Chemistry of Soil Processes, D. J. Green and M. H. B. Hayes, Eds., p. 593, John Wiley & Sons, New York, NY, USA.

Kaasalainen, M. and Yli-Halla, M.(2003). Use of sequential extraction to assess metal partitioning in soils. *Environmental Pollution*, 126 (2): 225–233.

Krishna, A.K. and Govil, P.K. (2004). Heavy metal contamination of soil around Pali Industrial Area, Rajasthan, India. *Environmental Geology*, 47: 38-44.

Kuo, S., Heilman, P. E, and Baker, A. S.(1983). Distribution and forms of copper, zinc, cadmium, iron, andmanganese in soils near a copper smelter. *Soil Science*, 135 (2): 101–109.

Lasat, M.M.(2000). Phytoextraction of metals from contaminated soil: a review of plant/soil/metal interaction and assessment of pertinent agronomic issues. *Journal of Hazardous Substances Research*, 2: 1–25.

Manahan, S.E.(2003). Toxicological Chemistry and Biochemistry, CRC Press, Limited Liability Company (LLC), 3rd edition.

Nagajyoti, P.C., Lee, K.D. and Sreekanth, T.V.M.(2010). Heavy metals, occurrence and toxicity for plants: A review. *Environ. Chem. Lett.*, 8: 199–216.

NSC, Lead Poisoning, National Safety Council. (2009). Resources/Documents/ Lead Poisoning.pdf. http://www.nsc.org/news

Pierzynski ,G. M., Sims J. T. and Vance G. F.(2000). Soils and Environmental Quality, CRC Press, London,UK, 2nd edition.

Reed, S. C., Crites, R. W., and Middlebrooks, E. J. (1995). Natural Systems for Waste Management and Treatment, McGraw-Hill, New York, NY, USA, 2nd edition.

Sayadi., M.H. and Rezaei, M.R.(2014). Impact of land use on the distribution of toxic metals in surface soils in Birjand city, Iran. *Proceedings of the International Academy of Ecology and Environmental Sciences,* 4(1): 18-29.

Scragg, A.(2006). Environmental Biotechnology, Oxford University Press, Oxford, UK, 2nd edition.

Shiowatana ,J., McLaren, R.G., Chanmekha, N. and Samphao, A. (2001). Fractionation of arsenic in soil by a continuous flow sequential extraction method. *Journal of Environmental Quality,* 30 (6): 1940–1949.

Smith, L.A., Means, J.L. and Chen, A. (1995). Remedial Options for Metals-Contaminated Sites, Lewis Publishers, Boca Raton, Fla, USA.

Sumner, M.E.(2000). Beneficial use of effluents, wastes, and biosolids. *Communications in Soil Science and Plant Analysis*, 31(11–14): pp. 1701–1715.

USEPA (1996). Recent Developments for In Situ Treatment of Metals contaminated Soils, U.S. Environmental Protection Agency, Office of Solid Waste and Emergency Response.

USDHHS (1999). Toxicological profile for lead, United States Department of Health and Human Services, Atlanta, Ga, USA.

Zojaji, F., Hassani, A.H. and Sayadi, M.H.(2014). Bioaccumulation of chromium by Zea mays in wastewater-irrigated soil. An experimental study. *Proceedings of the International Academy of Ecology and Environmental Sciences*, 4(2): 62-67.

Contamination of Heavy Metals in Agricultural Soils: Sources, Ecological Consequences, and Remediation Measures

Vinod Kumar, A. K. Chopra, Roushan K. Thakur and Jogendra Singh

Agroecology and Pollution Research Laboratory, Department of Zoology and Environmental Sciences, Gurukula Kangri University, Haridwar- 249404 (Uttarakhand), India

Abstract

Scattered literature is harnessed to critically review the possible sources, chemistry, potential biohazards and best available remedial strategies for a number of heavy metals (lead, chromium, arsenic, zinc, cadmium, copper, mercury, Fe and nickel) commonly found in contaminated soils. Soils polluted with heavy metals have become common across the globe due to increasing in geologic and anthropogenic activities. Plants growing on these soils show a reduction in growth, performance, and yield. Bioremediation is an effective method of treating heavy metal polluted soils. It is a widely accepted method that is mostly carried out *in situ*; hence it is suitable for the establishment/reestablishment of crops on treated soils. Microorganisms and plants employ different mechanisms for the bioremediation of polluted soils. However, the success of this approach largely depends on the species of organisms involved in the process. Remediation of heavy metal contaminated soils is necessary to reduce the associated risks, make the land resource available for agricultural production, enhance food security and scale down land tenure problems arising from changes in the land use pattern.

Keywords: Agriculture, Crop, Food chain, Heavy metals, Remediation, Soil contamination, Toxic,

Introduction

The soil is the principal constituent of earth ecosystem that comprises of a complicated mixture of organic matter, liquids, minerals, gases, and a diversity of organisms that sustain life (Huang *et al.*, 1998; Pathak *et al.*, 2010; Kumar and Chopra, 2016a,b). Heavy metals are natural constituents of the earth crust. A number of these elements are biologically essential and are introduced into aquatic enrichments by various anthropogenic activities (Omar and Al-Khashman, 2004; Kumar and Chopra, 2012a,b). Main anthropogenic sources of heavy metals exist in various industrial point sources e.g., present and former mining activities, foundries, smelters and diffuse sources such as piping, constituents of products, combustion of by-products, traffic industrial and human activities (Nilgun *et al.*, 2004; Pathak *et al.*, 2011).The composition and proportion of these constituents affect the physical, chemical, and biological properties of soil which in turn affect its agricultural suitability. Soil interfaces with the lithosphere, hydrosphere, atmosphere, and biosphere playing important role in nutrient and organic wastes recycling, inhabiting microflora that aid decomposition processes, provide a medium for plant growth and water storage. Soils continue to modify over time as a product of climate change, weathering, anthropogenic, and other biotic activities. It acts as a sink for all chemicals generated from anthropogenic and natural activities (Jonathan *et al.*, 2004; Kumar and Chopra, 2013a). The retention time of different substances in soil ecosystems is longer than in hydrosphere and atmosphere because contaminants accumulate quickly in soils and deplete at a slow rate. Heavy metals at trace levels present in natural water, air, dust, soil and sediments play an important role in human life (Kumar *et al.*, 2010). Soils are a critical environment where rock, air and water interface. Consequently, they are subjected to a number of pollutants due to different anthropogenic activities (Industrial, agricultural, transport etc.) (Facchinelli *et al.*, 2001; Jonathan *et al.*, 2004; Chopra *et al.*, 2013). The chemical composition of soil, particularly its metal content is environmentally important, because toxic metals concentration can reduce soil fertility, can increase input to the food chain, which leads to accumulatingtoxic metals in foodstuffs, and ultimately can endanger human health. Because of its environmental significance, studies to determine risk caused by metal levels in soil on human health and forest ecosystem have attracted attention in recent years (Denti *et al.*, 1998; Sandaa *et al.*, 1999; Arantzazu *et al.*, 2000 and Krzyztof *et al.*, 2004; Kumar and Chopra, 2015a).

Agricultural soils in this regard are greatly prone to anthropogenic substances which are used to enhance agricultural productivity. Wastewater irrigation, fossil fuel combustion, vehicle emission, mining/smelting activities, atmospheric deposition from municipal and industrial sectors, and application of fertilisers, pesticides, and sewage sludge have resulted in metal contamination of agricultural soils (Guvencc *et al.*, 2003 and Ali *et al.*, 2014). Increased rate of heavy metal addition in soils has also accelerated corresponding metal biogeochemical cycles. Along with metals, many other organic substances, i.e., dieldrin, aldrin, lindane, pentachlorobenzene, heptachlor, polychlorinated dibenzo-p-dioxins, endrin, mirex, hexachlorobenzene, endosulfans, dichlorodiphenyltrichloroethane (DDT), polychlorinated biphenyls (PCBs), toxaphene, chlordane, polychlorinated dibenzofurans, and chlordecone, are found in agricultural soils generated from a diverse set of municipal and industrial activities. Soil contamination/pollution by anthropogenic activities is an established phenomenon with reports which date back to 100 BC (Eney and Petzold 1987)

Heavy metals have relatively high density and are poisonous at extremely low concentrations. Elevated heavy metal concentrations in agricultural soils are particularly important due to their persistence, toxicity, long half-lives, and bioaccumulation potential (Ali *et al.* 2014). They are a natural constituent of the earth's crust, and many are necessary for normal metabolic functioning in plants, animals, and humans. In nature, their concentrations rarely exceed toxic levels unless intervened by any anthropogenic or natural activity. Their elevated concentration in surface soils and plants constitutes a general and recognised indication of environmental pollution (Panek 2000). Numerous studies to date have been conducted on the metal levels in soils and therein growing plants establishing environmental health concerns, i.e., food chain contamination (Chopra *et al.*, 2011, 2013; Kumar and Chopra, 2015a ; Kumar *et al.*, 2016a,b; Kumar and Chopra, 2016a,b).

India ranks second worldwide in farm output. Agriculture and allied sectors like forestry, logging and fishing accounted for 17% of the GDP and employed 49% of the total workforce in 2014. As the Indian economy has diversified and grown, agriculture's contribution to GDP has steadily declined from 1951 to 2011, yet it is still the largest employment source and a significant piece of the overall socio-economic development of India (Economic Survey 2010).

Heavy metals constitute an ill-defined group of inorganic chemical hazards, and those most commonly found at contaminated sites are lead (Pb), chromium (Cr), arsenic (As), zinc (Zn), cadmium (Cd), copper (Cu), mercury (Hg), and nickel (Ni) GWRTAC (1997). Soils are the major sink for heavy metals released into the environment by aforementioned anthropogenic activities and unlike organic contaminants which are oxidized to carbon (IV) oxide by microbial action, most metals do not undergo microbial or chemical degradation (Kirpichtchikova *et al.*, 2006; Kumar and Chopra, 2012a,b), and their total concentration in soils persists for a long time after their introduction (Adriano, 2003). Changes in their chemical forms (speciation) and bioavailability are, however, possible. The presence of toxic metals in soil can severely inhibit the biodegradation of organic contaminants (Maslin and Maier, 2000). Heavy metal contamination of soil may pose risks and hazards to humans and the ecosystem through: direct ingestion or contact with contaminated soil, the food chain (soil-plant-human or soil-plant-animal-human), drinking of contaminated ground water, reduction in food quality (safety and marketability) via phytotoxicity, reduction in land usability for agricultural production causing food insecurity, and land tenure problems (Ling *et al.*, 2007).

Heavy metals are known to affect crop quality/production, threatening human and livestock health through plant produce consumption. Such ecological risks associated with heavy metal contamination of agricultural soils are grave and urge remediation measures. Globally, food safety remains a major concern for food security. Plant metal uptake/contamination depends on the species type, metal loads in soils, bioavailability, and soil characteristics, i.e., pH, electrical conductivity, salinity, organic matter, texture, cation exchange capacity, sodium absorption ratio, and redox conditions. Atmospheric deposition can also be vital in metal deposition on plant surfaces via atmospheric dust thus contributing to crop contamination.

Soil reclamation strategies in India are often ill practised. Therefore, metal enrichment in agricultural soils continues indefinitely. Different lab-based studies have been continuously reported in literature for the metal removal from contaminated soils but unfortunately, are not replicated in polluted areas. Compared to the high-tech metal removal technologies, more focus has been on the bioremediation related techniques owing to their cost effectiveness and environment-friendly nature. Chemometric and geo-statistical approaches can be of momentous help in evaluating national metal loads and the identification of sources and hotspots that need to be reclaimed on top priority basis.

Sources of Heavy Metals in Agricultural Soil

Heavy metal sources in soils can be categorised into point metal sources and no point metal sources. Point metal sources indicate localised and discrete contamination source for heavy metals, whereas no point metal sources epitomise diffuse processes covering large areas (Kumar and Chopra, 2012b; Kumar *et al.,* 2016a). Examples for point metal sources in soil include mining/smelting and industrial and/or municipal activities; however, no point metal sources include adverse agricultural practices, fossil fuel burning, and atmospheric deposition. In point metal sources, usually excess heavy metal concentration is present in the immediate soils, whereas fairly less metal levels are found in soils having no point metal sources due to dilution effects. Some of the major sources of heavy metal in Indian soils are described hereunder.

Pesticide/Fertilizer Application

India is an agricultural country with a rapidly expanding population. To meet the growing population demands, excessive pesticides and fertilisers are utilised to enhance agricultural productivity. In irrigated areas, these chemicals (pesticides and fertilisers) are more excessively employed to boost agricultural production as compared to rain-fed areas.

Historically, agriculture was the first major human influence on the soil (Scragg, 2006; Kumar *et al.,* 2010). To grow and complete the lifecycle, plants must acquire not only macronutrients (N, P, K, S, Ca, and Mg), but also essential micronutrients. Some soils are deficient in the heavy metals (such as Co, Cu, Fe, Mn, Mo, Ni, and Zn) that are essential for healthy plant growth (Lasat, 2000), and crops may be supplied with these as an addition to the soil or as a foliar spray. Cereal crops grown on Cu-deficient soils are occasionally treated with Cu as an addition to the soil, and Mn may similarly be supplied to cereal and root crops. Large quantities of fertilisers are regularly added to soils in intensive farming systems to provide adequate N, P, and K for crop growth. The compounds used to supply these elements contain trace amounts of heavy metals (e.g., Cd and Pb) as impurities, which, after continued fertiliser, application may significantly increase their content in the soil (Jones and Jarvis, 1981).Metals, such as Cd and Pb, have no known physiological activity. Application of certain phosphatic fertilisers inadvertently adds Cd and other potentially toxic elements to the soil, including F, Hg, and Pb (Raven *et al.*, 1998; Chopra *et al.*, 2011, 2013).

Commonly employed fertilisers in India are nitrogen (N), phosphorus (P), and potassium (K), but along with them, boron (B)-, zinc (Zn)-, and sulphur (S)-based fertilisers are also used nowadays. Most abundantly employed fertilisers are nitrogenous followed by phosphate and potash. Small farmers who cannot afford commercial fertilisers due to high cost apply industrial wastes, sewage sludge, and animal manure in their agricultural fields to ensure good nutrient supply to growing crops. Fertilisers generated from the industrial and municipal wastes often contain appreciable amounts of heavy metals. Excess application of fertilisers in the agricultural soils not only results in heavy metal build-up in soils but also leads to other environmental problems, i.e., nutrient leaching from the agricultural lands to the underground aquifers and nearby water bodies causing eutrophication and environmental health problems. It is estimated that approximately 50 % of nutrients are lost from the fertiliser application sites (Kumar *et al.*, 2016a,b).

Wastewater irrigation

The application of municipal and industrial wastewater and related effluents to land dates back 400 years and now is a common practice in many parts of the world (Reed *et al.*, 1995). Worldwide, it is estimated that 20 million hectares of arable land are irrigated with waste water. cIn several Asian and African cities, studies suggest that agriculture based on wastewater irrigation accounts for 50 percent of the vegetable supply to urban areas (Bjhur, 2007). Farmers generally are not bothered about environmental benefits or hazards and are primarily interested in maximising their yields and profits. Although the metal concentrations in wastewater effluents are usually relatively low, long-term irrigation of land with such can eventually result in heavy metals accumulation in the soil (Pathak *et al.*, 2011; Chopra *et al.*, 2013).

Atmospheric Deposition

Atmospheric deposition from the industrial hubs, urban centres, and dense traffic areas is a common and diffuse source of heavy metals in surrounding soils. Global annual estimates of soil heavy metal accretion through atmospheric deposition are presented in Table 1.1 (Nriagu and Pacyna, 1988). Airborne sources of heavy metals from the aforementioned anthropogenic activities are in two forms, i.e., fugitive (dust) and stack/duct emissions (gases, air, or vapour streams) (Simonson, 1995). Stack/duct emissions usually transport heavy metals to distant areas, whereas fugitive emissions distribute metals to a considerably smaller area (Wuana and Okieimen, 2011; Chopra *et al.*, 2011).

Also, metal loads carried by the fugitive emissions are much less than the stack/dust emissions. Heavy metals in both emission forms after covering a distance from the originating source are deposited on land or water body. High metal loads, i.e., Zn, Pb, Cu, and Cd, are reported in the nearby agricultural areas of intense traffic hubs (Chopra *et al.*, 2011).

Table 1.1: Global estimates of soil heavy metals built-up through atmospheric fallouts (adopted from Nriagu and Pacyna, 1988).

S. N.	Heavy metals	10^6 kg/year
1.	Arsenic	8.4 – 18.0
2.	Cadmium	2.2 – 8.4
3.	Chromium	5.1 – 38
4.	Copper	14 – 36
5.	Mercury	0.63 – 4.3
6.	Manganese	7.4 – 46
7.	Molybdenum	0.55 – 4.0
8.	Nickel	11 – 37
9.	Lead	202 – 263
10.	Antimony	1.0 – 3.9
11.	Selenium	1.3 – 2.6
12.	Vanadium	3.2 – 21
13.	Zinc	49 – 135

Wind direction, wind speed, precipitation, and related climatic factors strongly influenced the rate and intensity of atmospheric deposition. Fossil fuel burning is an important contributor of heavy metals in Indian atmosphere. Prevailing power shortage at industrial and domestic levels has led to increase reliance on the fossil fuels to fulfill the energy demands in both sectors. This scenario has further worsened the air quality with respect to metal enrichment followed by its deposition on soils (Chopra *et al.*, 2011).

Ecological Risks Associated with Increasing Metals Loads in Agricultural Soils

Increased metal built-up in agricultural soils has led to serious ecological consequences in India, i.e., phytotoxicity, risks to soil dwelling organisms, food chain contamination, and public health problems. Food safety is currently the uprising concern which is at stake by various anthropogenic engagements (Kumar *et al.*, 2016a,b).

Effects of Heavy Metals on Soil and Plant Growth

The heavy metals that are available for plant uptake are those that are present as soluble components in the soil solution or those that are easily solubilized by root exudates (Blaylock and Huang, 2000). Although plants require certain heavy metals for their growth and upkeep, excessive amounts of these metals can become toxic to plants. The ability of plants to accumulate essential metals equally enables them to acquire other non-essentialmetals (Djingova and Kuleff,

2000). As metals cannot be broken down, when concentrations within the plant exceed optimal levels, they adversely affect the plant both directly and indirectly. The effect of heavy metal toxicity on the growth of plants varies according to the particular heavy metal involved in the process. Table 1.2 shows a summary of the toxic effects of specific metals on growth, biochemistry, and physiology of various plants. For metals such as Pb, Cd, Hg, and As which do not play any beneficial role in plant growth, adverse effects have been recorded at very low concentrations of these metals in the growth medium. Kibra (2008) recorded significant reduction in height of rice plants growing on a soil contaminated with 1mgHg/kg. Reduced tiller and panicle formation also occurred at this concentration of Hg in the soil. For Cd, reduction in shoot and root growth in wheat plants occurred when Cd in the soil solution was as low as 5mg/L (Ahamad *et al.*, 2012). Most of the reduction in growth parameters of plants growing on polluted soils can be attributed to reduced photosynthetic activities, plant mineral nutrition, and reduced activity of some enzymes.

Table 1.2: Toxic effects of heavy metals on some plant species.

Heavy Metal	Plant	Toxic effect on plant	Reference
As	Rice (*Oryza sativa*)	Reduction in seed germination; decrease in seedling height; reduced leaf area and dry matter production.	Abedin *et al.*, 2002
	Tomato (*Lycopersicon esculentum*)	Reduced fruit yield; decrease in leaf fresh weight	Barrachina *et al.*, 1995
	Canola (*Brassica napus*)	Stunted growth; chlorosis; wilting	Ahmad *et al.*, 2012
Cd	Wheat (*Triticum* sp.)	Reduction in seed germination; decrease in plant nutrient content; reduced shoot and root length	Yourtchi and Bayat, 2013
	Garlic (*Allium sativum*)	Reduced shoot growth; Cd accumulation	Jiang *et al.*, 2001
	Maize (*Zea mays*)	Reduced shoot growth; inhibition of root growth	Wang *et al.*, 2007
Co	Tomato (*Lycopersicon esculentum*)	Reduction in plant nutrient content	Jayakumar *et al.*, 2013
	Mung bean (*Vigna radiata*)	Reduction in antioxidant enzyme activities; decrease in plant sugar, starch, amino acids, and protein content	Jayakumar *et al.*, 2008
	Radish (*Raphanus sativus*)	Reduction in shoot length, root length, and total leafarea; decrease in chlorophyll content; reduction in plant nutrient content and antioxidant enzyme activity; decrease in plant sugar, amino acid, and protein content	Jayakumar *et al.*, 2007

Heavy Metal	Plant	Toxic effect on plant	Reference
Cr	Wheat (*Triticum*sp.)	Reduced shoot and root growth	Panda and Patra, 2000
	Tomato (*Lycopersicon esculentum*)	Decrease in plant nutrient acquisition	Moral *et al.*, 1996
	Onion (*Allium cepa*)	Inhibition of germination process; reduction of plant biomass	Nematshahi *et al.*, 2012
Cu	Bean (*Phaseolus vulgaris*)	Accumulation of Cu in plant roots; root malformation and reduction	Cook *et al.*, 1997
	Black bindweed (*Polygonum convolvulus*)	Plant mortality; reduced biomass and seed production	Kjær and Elmegaard, 1996
	Rhodes grass (*Chloris gayana*)	Root growth reduction	Sheldon and Menzies, 2005
Hg	Rice (*Oryza sativa*)	Decrease in plant height; reduced tiller and panicle formation; yield reduction; bioaccumulation in shoot and root of seedlings	Du *et al.*, 2005
	Tomato (*Lycopersicon esculentum*)	Reduction in germination percentage; reduced plant height; reduction in flowering and fruit weight; chlorosis	Shekar *et al.*, 2011
Mn	Broad bean (*Viciafaba*)	Mn accumulation shoot and root; reduction in shoot and root length; chlorosis	Arya and Roy, 2011
	Spearmint (*Mentha spicata*)	Decrease in chlorophyll a and carotenoid content; accumulation of Mn in plant roots	Asrar *et al.*, 2005
	Pea (*Pisumsativum*)	Reduction in chlorophylls a and b content; reduction in relative growth rate; reduced photosynthetic O2 evolution activity and photosystem II activity	Doncheva *et al.*, 2005
	Tomato (*Lycopersicon esculentum*)	Slower plant growth; decrease in chlorophyll concentration	Shenker *et al.*, 2004

Heavy Metal	Plant	Toxic effect on plant	Reference
Ni	Pigeon pea (*Cajanus cajan*)	Decrease in chlorophyll content and stomatalconductance; decreased enzyme activity which affected Calvin cycle and CO_2 fixation	Sheoran *et al.*, 1990
	Rye grass (*Lolium perenne*)	Reduction in plant nutrient acquisition; decrease in shoot yield; chlorosis	Khalid and Tinsley, 1980
	Wheat (*Triticum*sp.)	Reduction in plant nutrient acquisition	Barsukova and Gamzikova, 1999
	Rice (*Oryza sativa*)	Inhibition of root growth	Lin and Kao, 2005
Pb	Maize (*Zea mays*)	Reduction in germination percentage; suppressed growth; reduced plant biomass; decrease in plant protein content	Hussain *et al.*, 2013
	Portia tree (*Thespesia populnea*)	Reduction in number of leaves and leaf area; reduced plant height; decrease in plant biomass	Kabir *et al.*, 2009
	Oat (*Avena sativa*)	Inhibition of enzyme activity which affected CO_2 fixation	Moustakas *et al.*, 1994
Zn	Cluster bean (*Cyamopsis tetragonoloba*)	Reduction in germination percentage; reduced plant height and biomass; decrease in chlorophyll, carotenoid, sugar, starch, and amino acid content	Manivasagape-rumal *et al.*, 2011
	Pea (*Pisumsativum*)	Reduction in chlorophyll content; alteration in structure of chloroplast; reduction in photosystem II activity; reduced plant growth	Doncheva *et al.*, 2001
	Rye grass (*Lolium perenne*)	Accumulation of Zn in plant leaves; growth reduction; decrease in plant nutrient content; reduced efficiency of photosynthetic energy conversion	Bonnet *et al.*, 2000

It is important to note that certain plants are able to tolerate high concentration of heavy metals in their environment. Baker (1981) reported that these plants are able to tolerate these metals via 3 mechanisms, namely, (i) exclusion: restriction of metal transport and maintenance of a constant metal concentration in the shoot over a wide range of soil concentrations; (ii) inclusion: metal concentrations in the shoot reflecting those in the soil solution through a linear relationship; and (iii) bioaccumulation: accumulation of metals in the shoot and roots of plants at both low and high soil concentrations.

Food Chain Contamination and Human Health Concerns

To date, various researchers have documented phytotoxicity, food chain contamination, and mounting human health concerns from different agroecological zones of India. Singh *et al.*, (2010) described metal contamination of vegetables and associated human health risk from different regions of a dry tropical area of India. The risk to human health by heavy metals (Cd, Cu, Pb, Zn, Ni and Cr) through the intake of locally grown vegetables, cereal crops and milk from wastewater irrigated site. Milk is not directly contaminated due to wastewater irrigation but is an important route of food chain transfer of heavy metals from grass to animals. Singh *et al.* (2010) reported concentrations of Cd, Pb and Ni have crossed the safe limits for human consumption in all the vegetables in Varanasi region.

Mitigation Measures

Agricultural soil pollution is has become critical in India. Soil reclamation strategies are not potentially implemented as in developed countries. Research on exploring prospective methods regarding soil remediation is in the process. Biological methods (bioremediation) hold a special place in soil reclamation strategies and are more extensively researched as compared to physical, chemical, or mechanical methods. Bioremediation technology is not only cost-effective and environment-friendly, but it is also the best alternative to other conventional treatments including immobilisation, volatilization, and incineration (Iram *et al.*, 2009). Recent soil reclamation research in India has featured fungal, plant, and bacterial usage in soil bioremediation.

Mycoremediation

Mycoremediation is a process of using fungi to return an environment (usually soil) contaminated by pollutants to a less contaminated state. It means using various strains of fungi to clean as a radionuclide. Mycoremediation also held promise for removing heavy metals from the land by channelling them to the fruit bodies for removal (Okhuoya, 2011). One of the primary roles of fungi in the ecosystem is decomposition, which is performed by the mycelium. The mycelium secretes extracellular enzymes and acids that breakdown lignin and cellulose, the two main building blocks of plant fibre. The key to mycoremediation is determining the right fungi species to target a specific pollutant. Asiriuwa *et al.* (2013) studied mycoremediation technique was used to assess the bioaccumulation potential of heavy metal (Cd, Zn, Cu, Pd) by mushroom from heavy metal contaminated soils. Mycoremediation is a form of bioremediation in which various strains of fungi are used to decontaminate contaminated environment. Results obtained from the study revealed that mushroom can bioaccumulate heavy metal from metal contaminated soil

Phytoremediation

Phytoremediation, also called green remediation, botanoremediation, agroremediation, or vegetative remediation, can be defined as an *in situ* remediation strategy that uses vegetation and associated microbiota, soil amendments, and agronomic techniques to remove, contain, or render environmental contaminants harmless (Helmisaari *et al.*, 2007; Raymond and Felix, 2011).

The idea of using metal-accumulating plants to remove heavy metals and other compounds was first introduced in 1983, but the concept has actually been implemented for the past 300 years on wastewater discharges (Henry, 2000). Plants may break down or degrade organic pollutants or remove and stabilise metal contaminants. The methods used to phytoremediate metal contaminants are slightly different from those used to remediate sites polluted with organic contaminants. As it is a relatively new technology, phytoremediation is still mostly in its testing stages and as such has not been used in many places as a full-scale application. However, it has been tested successfully in many places around the world for many different contaminants. Phytoremediation is energy efficient, aesthetically pleasing method of remediating sites with low to moderate levels of contamination, and it can be used in conjunction with other more traditional remedial methods as a finishing step to the remedial process (Kumar and Chopra, 2016a).

The advantages of phytoremediation compared with classical remediation are that (i) it is more economically viable using the same tools and supplies as agriculture, (ii) it is less disruptive to the environment and does not involve waiting for new plant communities to recolonize the site, (iii) disposal sites are not needed, (iv) it is more likely to be accepted by the public as it is more aesthetically pleasing then traditional methods, (v) it avoids excavation and transport of polluted media thus reducing the risk of spreading the contamination, and (vi) it has the potential to treat sites polluted with more than one type of pollutant. The disadvantages are as follow (i) it is dependent on the growing conditions required by the plant (i.e., climate, geology, altitude, and temperature), (ii) large-scale operations require access to agricultural equipment and knowledge, (iii) success is dependent on the tolerance of the plant to the pollutant, (iv) contaminants collected in senescing tissues may be released back into the environment in autumn, (v) contaminants may be collected in woody tissues used as fuel, (vi) time taken to remediate sites far exceeds that of other technologies, (vii) contaminant solubility may be increased leading to greater environmental damage and the possibility of leaching. Potentially useful phytoremediation technologies for remediation of heavy metal-contaminated soils include phytoextraction (phytoaccumulation), phytostabilization, and phytofiltration (Raymond and Felix, 2011).

Conclusion

The increased knowledge regarding the deleterious impacts of heavy metals has led to the basic legislation in India to combat mounting metal levels in different environmental compartments. Still, however, soil background metal levels (preindustrial) are not established in the agricultural, urban, industrial, or rural soils of India. Similarly, soil metal standards are not available at the federal and/or provincial level to maintain healthy soil ecosystems. Based on the recent studies from irrigated and rain-fed areas of India the allowable standard/limits set by international agencies leading to food chain contamination. Public and livestock health is under serious toxicological threat where high metal levels are recorded both in the soils and the food crops. Such adverse soil ecological conditions urge reclamation procedures that befit India's agricultural environments. Bioremediation (i.e., mycoremediation, phytoremediation, and bacterial bioremediation) has been extensively studied by scientists in India for devising metal abatement strategies in the contaminated soils. Legislation at the federal and provincial level needs to be revised and strengthened to conserve this precious resource. Farmer's awareness, good agricultural practices, improved performance of the agricultural extension departments, and strict implementation of legislation regarding metal diminution can bring reduction in the soil metal pollution loads.

References

Arantzazu, U.,Vega, M. and Angul, E. (2000). Deriving ecological risk based soil quality values in the Barque country. *The Science of the Total Environment*, 247: 279–284.

Abedin, M. J., Cotter-Howells, J., and Meharg, A. A. (2002). Arsenic uptake and accumulation in rice (*Oryza sativa* L.) irrigated with contaminated water. *Plant and Soil*, 240(2): 311–319.

Adriano, D. C. (2003). *Trace Elements in Terrestrial Environments: Biogeochemistry, Bioavailability and Risks of Metals*, Springer, New York, NY, USA, 2nd edition.

Asrar, Z., Khavari-Nejad, R. A. and Heidari, H. (2005). Excess manganese effects on pigments of *Mentha spicata* at flowering stage. *Archives of Agronomy and Soil Science*, 51 (1): 101–107.

Arya, S. K. and Roy B. K. (2011). Manganese induced changes in growth, chlorophyll content and antioxidants activity in seedlings of broad bean (*Vicia faba* L.). *Journal of Environmental Biology*, 32 (6): 707–711.

Ahmad, I., Akhtar, M. J., Zahir, Z. A. and Jamil, A. (2012). Effect of cadmium on seed germination and seedling growth of four wheat (*Triticum aestivum* L.) cultivars. *Pakistan Journal of Botany*, 44 (5): 1569–1574.

Asiriuwa, O.D., Ikhuoria, J.U. and Ilori, E.G. (2013). Myco-Remediation Potential of Heavy Metals from Contaminated Soil. *Bull. Env. Pharmacol. Life Sci.*, 2 (5): 16-22.

Ali, Z., Malik, R.N., Shinwari, Z.K. and Qadir, A. (2014). Enrichment, risk assessment, and statistical apportionment of heavy metals in tannery-affected areas. *Int. J. Environ. Sci. Tech.*, 1–14. Available from: http://dx.doi.org/10.1007/s13762-013-0428-4.

Baker, A. J. M. (1981). Accumulators and excluders strategies in the response of plants to heavy metals. *Journal of Plant Nutrition*, 3: 643–654.

Barrachina, A. C., Carbonell, F. B., and Beneyto J. M. (1995). Arsenic uptake, distribution, and accumulation in tomato plants: effect of arsenite on plant growth and yield. *Journal of Plant Nutrition*, 18 (6): 1237–1250.

Barsukova, V. S. and Gamzikova O. I. (1999). Effects of nickel surplus on the element content in wheat varieties contrasting in Ni resistance. *Agrokhimiya*, 1: 80–85.

Bonnet, M., Camares, O., and Veisseire, P., (2000). Effects of zinc and influence of *Acremonium loliion* growth parameters, chlorophyll a fluorescence and antioxidant enzyme activities of rye grass (*Lolium perenne* L. cvApollo). *Journal of Experimental Botany*, 51 (346): 945–953.

Blaylock, M. J. and Huang, J.W., (2000). Phytoextraction of metals in *Phytoremediation of Toxic Metals: Using Plants to Clean up theEnvironment*, I. Raskin and B.D. Ensley, Eds., pp. 53–70, Wiley, New York, NY, USA.

Bjuhr J., (2007). *Trace Metals in Soils Irrigated with Waste Water in a Periurban Area Downstream Hanoi City, Vietnam, Seminar Paper*, Institutionen for markvetenskap, Sveriges lantbruksuniversitet(SLU), Uppsala, Sweden.

Chopra, A.K., Srivastava, S. and Kumar, V. (2011). Comparative study on agro-potentiality of Paper mill effluent and synthetic nutrient (DAP) on *Vigna unguiculata* L. (Walp) Cowpea. *Journal of Chemical and Pharmaceutical Research*, 3(5):151-165.

Chopra, A.K., Srivastava, S., Kumar, V. and Pathak, C. (2013). Agro-potentiality of distillery effluent on soil and agronomical characteristics of *Abelmoschus esculentus* L. (Okra). *Environmental Monitoring and Assessment*, 185: 6635-6644.

Cook, C.M., Kostidou, A., Vardaka, V, and Lanaras, V. (1997). Effects of copper on the growth, photosynthesis and nutrient concentrations of *Phaseolus* plants. *Photosynthetica*, 34 (2): 179–193.

Denti, B., Cocucci, S.M., and Di Givolamo, F. (1998). Environmental pollution and forest stress: a multidisciplinary approach study on alpine forest ecosystems. *Chemosphere*, 36: 1049–1054.

Djingova, R. and Kuleff, I. (2000). "Instrumental techniques for trace analysis," in *Trace Elements: Their Distribution and Effects inthe Environment*, J. P. Vernet, Ed., Elsevier, London,UK.

Doncheva, S., Stoynova, Z. and Velikova, V. (2001). Influence of succinate on zinc toxicity of pea plants," *Journal of Plant Nutrition*, 24 (6): 789–804.

Du, X., Zhu, Y.G., Liu, W.J. and Zhao, X.S. (2005). Uptake ofmercury (Hg) by seedlings of rice (*Oryza sativa* L.) grown in solution culture and interactions with arsenate uptake," *Environmentaland Experimental Botany*, 54 (1): 1–7.

Doncheva, S., Georgieva, K., Vassileva, V., Stoyanova, Z., Popov, N., and Ignatov G., (2005). Effects of succinate on manganese toxicity in pea plants. *Journal of Plant Nutrition*, 28 (1): 47–62.

Eney A, Petzold D (1987). The problem of acid rain: an overview. *Environmentalist* (2):95-103.

Facchinelli, A., Sacchi, E. and Malleri, L. (2001). Multivariate statistical and GIS-based approach to identify heavy metal sources in soils. *Environmental Pollution*, 114: 313–324.

GWRTAC (1997). "Remediation of metals-contaminated soils and groundwater," Tech. Rep. TE-97-01, GWRTAC, Pittsburgh, Pa, USA, GWRTAC-E Series.

Guvencc, N, Alagha, O. and Tuncel, G. (2003). Investigation of soil multi-element composition in Antalya, Turkey. *Environ Int.*, 29(5):631–640.

Huang, P., Adriano, D., Chlopecka, A. and Kaplan, D. (1998). Role of soil chemistry in soil remediation and ecosystem conservation. In: Huang PM (ed.), Soil chemistry and ecosystem health. Madison, WI: Soil Science Society of America. Available from:http://dx.doi.org/10.2136/sssaspecpub52.c13.

Henry R. J., (2000). *An Overview of the Phytoremediation of Lead andMercury*, United States Environmental Protection AgencyOffice of Solid Waste and Emergency Response Technology Innovation office,Washington, DC, USA.

Helmisaari, H. S., Salemaa M., Derome J., Kiikkil, O., Uhlig, C. and Nieminen T. M., (2007). Remediation of heavy metalcontaminated forest soil using recycled organic matter and native woody plants. *Journal of Environmental Quality*, 36 (4): 1145–1153.

Hussain, A, Abbas, N. and Arshad, F. (2013). "Effects of diverse doses of lead (Pb) on different growth attributes of *Zea mays* L. *Agricultural Sciences*, 4 (5): 262–265.

Iram, S., Ahmad, I., Javed, B., Yaqoob, S., Akhtar, K., Kazmi, M.R. and Zaman, B. (2009) Fungal tolerance to heavy metals. *Pak. J. Bot.*, 41(5):2583–2594.

Jones, L. H. P and Jarvis, S. C. (1981). "The fate of heavy metals," in *The Chemistry of Soil Processes*, D. J. Green and M. H. B. Hayes, Eds., p. 593, JohnWiley & Sons, New York, NY, USA.

Jiang, W., Liu, D. and Hou, W. (2001). Hyperaccumulation of cadm ium by roots, bulbs and shoots of garlic," *Bioresource Technology*,76 (1): 9–13.

Jonathan, M., Ram-Mohan, V. and Srinivasalu, S. (2004). Geochemical variations of major and trace elements in recent sediments, off the Gulf of Mannar, the southeast coast of India. *Environ. Geol.*, 45(4):466–480.

Jayakumar, K., Jaleel, C. A. and Vijayarengan, P. (2007). Changes in growth, biochemical constituents, and antioxidant potentials in radish (*Raphanus sativus* L.) under cobalt stress," *TurkishJournal of Biology*, 31(3): 127–136.

Jayakumar, K., Jaleel, C. A. and Azooz, M. M. (2008). Phytochemical changes in green gram (*Vigna radiata*) under cobalt stress. *Global Journal of Molecular Sciences*, 3 (2): 46–49.

Jayakumar, K., Rajesh, M., Baskaran, L. and Vijayarengan P., (2013). Changes in nutritional metabolism of tomato (*Lycopersicon esculantum* Mill.) plants exposed to increasing concentration of cobalt chloride. *International Journal of Food Nutrition and Safety*, 4 (2): 62–69.

Kabir, M., Iqbal, M. Z., and Shafiq, M., (2009). Effects of lead on seedling growth of *Thespesia populnea* L.. *Advances in Environmental Biology*, 3 (2): 184–190.

Khalid, B. Y. and Tinsley, J. (1980). Some effects of nickel toxicity on rye grass, *Plant and Soil*, vol. 55(1): 139–144.

Kjær, C. and Elmegaard, N. (1996). "Effects of copper sulfate on black bindweed (*Polygonum convolvulus* L.). *Ecotoxicology and Environmental Safety*, 33 (2): 110–117.

Krzyztof, Loska, Danutta, Wiechua, and Irena, Korus. (2004). Metal contamination of farming soils affected by industry. *Environment International*, 30(2): 159–165.

Kirpichtchikova T. A., Manceau A., Spadini L., Panfili F., Marcus M. A., and Jacquet T. (2006). "Speciation and solubility of heavy metals in contaminated soil using X-ray microfluorescence, EXAFS spectroscopy, chemical extraction, and thermodynamic modeling," *Geochimica et Cosmochimica Acta*, vol. 70, no. 9, pp. 2163–2190.

KibraM. G., (2008). "Effects of mercury on some growth parameters of rice (*Oryza sativa* L.)," *Soil & Environment*, vol. 27, no. 1, pp. 23–28.

Kumar, V., Chopra A.K., Pathak, C. and Pathak, S. (2010). Agro-potentiality of Paper Mill Effluent on the characteristics of *Trigonella foenum-graecum* L. (Fenugreek) *New York Science Journal*, 3(5): 68-77.

Kumar, V. and Chopra A.K. (2012a). Fertigation effect of distillery effluent on agronomical practices of *Trigonella foenum-gruecum* L. (Fenugreek). *Environmental Monitoring and Assessment*, 184:1207-1219.

Kumar, V. and Chopra A.K. (2012b). Effects of paper mill effluent irrigation on agronomical characteristics of *Vigna radiata* (L.) in two different seasons. *Communications in Soil Science and Plant Analysis*, 43(16).2142-2166.

Kumar, V. and Chopra, A.K. (2013a). Ferti-irrigational effect of paper mill effluent on agronomical characteristics of *Abelmoschus esculentus* L. (Okra). *Pakistan Journal of Biological Sciences*, 16 (22): 1426-1437.

Kumar, V. and Chopra, A.K. (2013b). Enrichment and translocation of heavy metals in soil and *Vicia faba* L. (Faba bean) after fertigation with distillery effluent. *International Journal of Agricultural Policy and Research*, 1(5): 131-141.

Kumar, V. and Chopra, A.K. (2014a). Ferti-irrigational impact of sugar mill effluent on agronomical characteristics of *Phaseolus vulgaris* (L.) in two seasons. *Environmental Monitoring and Assessment*, 186:7877–7892.

Kumar, V. and Chopra, A.K. (2014b). Ferti-irrigational response of hybrid cultivar of Indian mustard (*Brassica juncea* L.) to distillery effluent in two seasons. *Analytical Chemistry Letters*, 4(3): 190-206.

Kumar, V. and Chopra, A.K. (2015a). Fertigation with agro-residue based paper mill effluent on a high yield spinach variety. *International Journal of Vegetable Science*, 21(1): 69-97.

Kumar, V. and Chopra, A.K. (2015b). Heavy metals accumulation in soil and agricultural crops grown in the Province of Asahi India Glass Ltd., Haridwar (Uttarakhand), India. *Advances in Crop Science and Technology*, 4: 203.

Kumar, V. and Chopra, A.K. and Srivastava S. (2016a). Assessment of heavy metals in spinach (*Spinacia oleracea* L.) grown in sewage sludge amended soil. *Communications in Soil Science and Plant Analysis*, 47(2): 221-236.

Kumar, V., Gautam, P., Singh, J. and Thakur, R.K. (2016b). Assessment of phycoremediation efficiency of *Spirogyra Sp.* using sugar mill effluent. *International Journal of Environment, Agriculture and Biotechnology*, 1(1): 54-62.

Kumar, V. and Chopra, A.K. (2016a). Agronomical performance of high yielding cultivar of eggplant (*Solanum melongena* L.) grown in sewage sludge amended soil. *Research in Agriculture*, 1(1): 1-24.

Kumar, V. and Chopra, A.K. (2016b). Effects of sugarcane pressmud on agronomical performance of eggplant (*Solanum melongena* L.) grown in sewage sludge amended soil in field conditions. *International Journal of Recycling of Organic Waste in Agriculture*, 5: 149-162.

Lasat M. M (2000). "Phytoextraction of metals from contaminated soil: a review of plant/soil/metal interaction and assessment of pertinent agronomic issues," *Journal of Hazardous SubstancesResearch*, vol. 2, pp. 1–25.

Lin Y.-C and Kao C.-H., (2005). "Nickel toxicity of rice seedlings: Cell wall peroxidase, lignin, and NiSO4-inhibited root growth," *Crop, Environment Bioinformatics*, vol. 2, pp. 131–136.

LingW., ShenQ., GaoY., GuX., and YangZ. (2007). "Use of bentonite to control the release of copper from contaminated soils," *Australian Journal of Soil Research*, vol. 45, no. 8, pp. 618–623.

Moustakas M., Lanaras T., Symeonidis L., and Karataglis S., (1994). "Growth and some photosynthetic characteristics of field grown Avena sativa under copper and lead stress," *Photosynthetica*, vol. 30, no. 3, pp. 389–396.

Moral R., Gomez I., Pedreno J. N., and Mataix J., (1996). "Absorption of Cr and effects on micronutrient content in tomato plant (*Lycopersicumesculentum*M.)," *Agrochimica*, vol. 40, no. 2-3, pp. 132 -138.

Maslin P. and Maier R. M.(2000). "Rhamnolipid-enhanced mineralization of phenanthrene in organic-metal co-contaminated soils," *Bioremediation Journal,* vol. 4, no. 4, pp. 295–308.

Manivasagaperumal R., Balamurugan S., Thiyagarajan G., and Sekar J., (2011). "Effect of zinc on germination, seedling growth and biochemical content of cluster bean (*Cyamopsis tetragonoloba* (L.) Taub)," *Current Botany,* vol. 2, no. 5, pp. 11–15.

Nriagu J, Pacyna J (1988). Quantitative assessment of worldwide contamination of air, water and soils by trace metals. Nature 333(6169):134–139.

Nilgun, Guvenc, Omar, Alagha, & Gurdal, Tencel. (2004). Investigation of soil multi-element composition in Antalya, Turkry. *Environmental International,* 29, 631–640.

Nematshahi N., Lahouti M., and Ganjeali A., (2012). "Accumulation of chromium and its effect on growth of (*Allium cepa* cv. Hybrid)," *European Journal of Experimental Biology,* vol. 2, no. 4, pp. 969–974.

Omar, A., & Al-Khashman. (2004). Heavy metal distribution in dust, street dust and soils from the work place in Karak Industrial Estate, Jordan. *Atmospheric Pollution,* 38, 6803–6812.

Okhuoya J.A., (2011). Mushrooms: what they are and what they do. Inaugural Lecture Series 114, University of Benin.

Panek E (2000) Metale sladowe w glebach i wybranych gatunkach roslin obszaru polskiej czesci Karpat. IGSMiE PAN, Krakow, Poland.

Panda S. K. and Patra H. K., (2000). "Nitrate and ammonium ions effect on the chromium toxicity in developing wheat seedlings,"*Proceedings of the National Academy of Sciences, India,* vol. 70, pp. 75–80.

Pathak, C., Chopra, A.K., Kumar, V., and Srivastava, S. (2010). Heavy metals contamination in wastewater irrigated agricultural soil near Bindal River, Dehradun, India. *Pollution Research,* 29 (4): 33-37.

Pathak, C., Chopra A. K., Kumar, V., and Sharma, S. (2011). Effect of sewage-water irrigation on physico-chemical parameters with special reference to heavy metals in agricultural soil of Haridwar city. *Journal of Applied and Natural Science,* 3 (1): 108-113.

Reed, S.C., Crites. W., and Middlebrooks E. J (1995). *Natural Systems forWasteManagement and Treatment,* McGraw-Hill,New York, NY, USA, 2nd edition.

RavenP. H,, BergL. R., and Johnson G. B (1998). *Environment,* Saunders College Publishing, New York, NY, USA, 2nd edition.

Raymond A.W and Felix E. O, (2011). HeavyMetals in Contaminated Soils: A Review of Sources, Chemistry, Risks and Best Available Strategies for Remediation, *International Scholarly Research Network.* doi:10.5402/2011/402647

Sheoran, S., Singal, H. R., and Singh R., (1990). "Effect of cadmium and nickel on photosynthesis and the enzymes of the photosynthetic carbon reduction cycle in pigeonpea (*Cajanus cajan* L.),"*Photosynthesis Research*, vol. 23, no. 3, pp. 345–351.

Simonson R (1995). Airborne dust and its significance to soils. Geoderma 65(1–2):1–43

Sandaa, R.A., Enger, O., & Torsvik, V. (1999). Abundance and diversity of archae in heavy-metal-contaminated soils. *Applied Environmental Microbiology, 65*, 3293–3297.

Shenker M., Plessner O. E., and Tel-Or E., (2004). "Manganese nutrition effects on tomato growth, chlorophyll concentration, and superoxide dismutase activity," *Journal of Plant Physiology*, vol. 161, no. 2, pp. 197–202.

Sheldon A. R. and Menzies N.W., (2005). "The effect of copper toxicity on the growth and root morphology of Rhodes grass (*Chlorisgayana* Knuth.) in resin buffered solution culture," *Plant and Soil*, vol. 278, no. 1-2, pp. 341–349.

Scragg A., (2006). *Environmental Biotechnology*, Oxford University Press, Oxford, UK, 2nd edition.

Singh, A., Sharma, R.K, Agrawalm and Marshall, F. M, (2010) *Tropical Ecology* **51**(2S): 375-387.

Shekar, C. H. C., Sammaiah D., Shasthree T., and Reddy K. J., (2011). "Effect of mercury on tomato growth and yield attributes,"*International Journal of Pharma and Bio Sciences*, vol. 2, no. 2, pp. B358–B364, 2011.

Wang M., Zou J., Duan X., Jiang W., and Liu D., (2007). "Cadmium accumulation and its effects on metal uptake in maize (*Zea mays* L.). *Bioresource Technology*, 98 (1): 82–88, 2007.

Wuana, R. and Okieimen, F. (2011). Heavy metals in contaminated soils: a review of sources, chemistry, risks and best available strategies for remediation. ISRN Ecol (2011), Article ID 402647, 20 pages, Available from:http://dx.doi.org/10.5402/2011/402647

Yourtchi M. S. and Bayat H. R., (2013). "Effect of cadmium toxicity on growth, cadmium accumulation and macronutrient content of durum wheat (Dena CV.)," *International Journal of Agricultureand Crop Sciences*. 6 (15): 1099–1103.

An Efficient Approach towards the Bioremediation of Heavy Metal Pollution from Soil and Aquatic Environment: An Overview

Vikas Kumar[1], Anjali Tiwari[2] and Amar Prakash Garg[3]

[1]School of Agricultural Sciences, Jaipur National University, Jaipur, Rajasthan (India)
[2]Department of Agriculture & Forestry, Dr. K.N. Modi University, Newai, Rajasthan (India)
[3]Jaipur National University, Jaipur, Rajasthan, India

Abstract

This review article provide an overall view of bioremediation which is cost-effective and environment eco-friendly technique for converting the toxic, recalcitrant pollutants into environmentally benign products through the action of various biological treatments. Plant-based bioremediation is still at a nascent stage, but it has shown immense promise as an alternative approach for cleaning up, especially the biosphere component of the environment. There are several mechanisms of bioremediation which including biosorption, metal-microbe interactions, bioaccumulation, biomineralisation, biotransformation and bioleaching.

Keywords: Bioremediation, heavy metals, biosorption, **metal-microbe interactions**, bioaccumulation, biomineralisation, bioformulation, bioleaching and bio-augmentation.

Introduction

Due to tremendous increase in population, urbanization and industrialization, extra dosage of fertilizer and pesticide application for intensive agriculture, many environmental issues such as solid waste and residue management, wastewater treatment and management, industrial and hazardous waste treatment, air pollution and management, aerospace and atomic energy installations, mining, surface finishing, electroplating and electric appliance manufacturing are imparting a huge negative impact on agricultural land and the water table underneath the existing soil profile since last two decades. Bioremediation has been the target of extensive studies as a clean-up technology for organic pollutant removal from contaminated soils. Similarly, heavy metals pollution of the environment has become a global catastrophe causes adverse effects on flora, fauna and groundwater contamination through leaching. Environmental pollution with metals, semi-metals and organic contaminants is a serious global problem, with heavy metals being one of the most dangerous pollutants (Xiezhi *et al.*, 2005). Heavy metals constitute an ill-defined group of inorganic chemical hazards, and those most commonly found at contaminated sites are lead (Pb), chromium (Cr), arsenic (As), zinc (Zn), cadmium (Cd), copper (Cu), mercury (Hg), and nickel (Ni) (GWRTAC, 1997). McLaughlin *et al.* (2000a,b) and Ling *et al.* (2007) revealed that heavy metal contamination of soil may pose risks and hazards to humans and the ecosystem through: direct ingestion or contact with contaminated soil, the food chain (soil-plant-human or soil-plant-animal human), drinking of contaminated ground water, reduction in food quality (safety and marketability) via phytotoxicity, reduction in land usability for agricultural production causing food insecurity, and land tenure problems. Therefore, the search for alternative methods to restore polluted sites in a less expensive, less labor-intensive, safe and environmentally friendly way is required against these environmental issues without compromising the greater economic development has become extremely important for developing country like India. Indeed, bioremediation is one of the possibility technology offers an environment friendly, cost effective, sustainable technology way of removing, altering, degrading, immobilizing or detoxifying various chemicals from the environment through the action of bacteria, fungi and plant offers promising technological avenues. Moreover, biological processes benefit from a high public acceptance together with a growing interest and awareness. Considering the hydrophobic properties of polycyclic aromatic hydrocarbons (PAHs), the success of soil bioremediation could be limited by many physicochemical and biological factors, but the most important is the bio availability of contaminants (Picado *et al.*, 2001; Sayara *et al.*, 2011). Although, recent scientific data and biotechnological showed the widespread discussions and applications pertaining to both the removal of toxic contaminants and recovery of valuable resources (nutrients and energy) from contaminated environments, e.g., from wastewater streams (Morel *et al.*, 2002; Verma *et al.*, 2006). However, bioremediation has potential to biodegradation, cost effective, cleans the polluted environment through the growth of microbial activities (algae, fungi, yeast or bacteria) and it have proved to be potential mental biosorbents, due to metal sequestering properties and can decrease the concentration of heavy metal ions in solution (Kamaludeen *et al.*, 2003; Tang *et al.*, 2007; Mishra and Malik, 2014; Balaji *et al.*, 2014; Uqab *et al.*, 2016) and known as "Eco bio technology".

When microorganisms are exposed to new substrates or growth conditions, they are able to synthesize new enzymes to yield energy and nutrients from different substrates or under new growth conditions after an acclimation period. On the other words, microorganisms interact chemically and physically with the hazardous organic compounds and result in their structural changes or complete degradation (Raymond *et al.*, 2001). This is because microorganism enzymes can be directly modified, modified by site-specific mutagenesis or modified in the development of biocatalytic (new enzymes) processes (Lim, 2015). These microbes transform the pesticides and other harmful toxic compounds into non-toxic substances and finally into carbon dioxide and water. Complex and effective metabolic pathways with unique set of enzymes are involved during the degradation by microorganisms.

Instance, fungi transform pesticides and other xenobiotic compounds by changing the structure of molecules and rendering them into nontoxic forms. These nontoxic forms of compounds are then completely degraded (Raymond *et al.*, 2001). For examples-Acinetobacter sps., Bascillus sps., Burkholderia sps., *Dehalospirillum multivorans*, Herbaspirillum sps., Nocardioides sps., Novosphingobium sps., *Methylobacterium populi*, Pseudomonas sps., Rhodococcus sps., Sphingobium sps., Sphingomonas sps., are bacteria which effective strains for degrading organic compounds (McGuinness and Dowling, 2009, Chaudhary and Kim, 2016). Several researchers (Dhankhar and Hooda, 2011; Mani and Kumar, 2013; Merugu *et al.*, 2014; Fonti *et al.*, 2015) noticed that physicochemical approaches are gaining increasing prominence in the remediation of a variety of environmental matrices because they are cost effective, environmentally friendly and they are associated with fewer side effects. Microbial degradation is an important mechanism controlling the fate of pesticides in soils, and is generally considered to be desirable both from an environmental as well agricultural perspective. In general, bioremediation is the use of microorganisms to degrade environmental contaminants (pesticides, polyaromatic hydrocarbons etc.) into less toxic forms or compounds. Erguven *et al.* (2016) reported that pesticides in agriculture acquired great importance in the past due to their pest control features, but now the focus is on their potential impact to human health and the environment (Khosravi *et al.*, 2009). Their existence in soil result to environmental stresses that can lead to reduction of plant growth (Khosravi *et al.*, 2009; Mohsenzade *et al.*, 2012). Thedegradation of pesticides is facilitated by both biotic and abiotic factors, including chemical, sunlight and microbial agents. Among these factors, biodegradation is the most commonly used method for converting synthetic chemicals into inorganic products (Basseyand Grigson, 2011). Gaur *et al.* (2014) examined that bioremediation employs diverse microbes for degradation or treatment of xenobiotic compounds, aromatic hydrocarbons, volatile organic compounds, pesticides, herbicides, heavy metals, radionuclides, crude oil, jet fuels, petroleum products and explosives. Dua *et al.* (2002) mentioned that important parameters for bioremediation are nature of the pollutants; soil structure and hydrogeology (movement of pollutants through soil and ground water) and the nutritional state and microbial composition of the site.

Principles of Bioremediation: Bioremediation is to detoxify contaminants from a given environment using microorganisms. On the other way, the environment friendly process of detoxification of harmful pollutants from soil, water and air using microorganisms (Mitra and Mukhopadhyay, 2016).Several investigators showed the excellent performance for the toxicity reduction of contaminants using physicochemical, thermal and Advanced Oxidation Processes (AOPs) including ozone, ozone/UV, ozone/H_2O_2 and electron beam/gamma irradiation.

Mechanisms of Bioremediation: Microorganisms may have to potential of heavy metal-contaminated soil and can easily convert heavy metals into non-toxic forms by using chemicals for their growth and development and also capable of dissolving metals and reducing or oxidizing transition metals. Bioremediation processes, microorganisms mineralize theorganic contaminants to end-products such as carbon dioxide and water, or to metabolic intermediates which are used as primary substrates for cell growth. Two-way defense activities happening in micro organisms *viz.* production of degradative enzymes for the target pollutants as well as resistance to relevant heavy metals. There are several mechanisms of bioremediation which including biosorption, metal-microbe interactions, bio accumulation, biomineralisation, biotransformation and bioleaching. Different methods by whichmicrobes restore the environment are oxidizing, binding, immobilizing, volatizing and transformation of heavy metals. Sikkema *et al.* (1995) reported that many contaminants are organic solvents which disrupt membranes, but cells may develop defense mechanisms including formation of outer cell-membrane-protective material, often hydrophobic or solvent efflux pumps. For instance, plasmid-encoded and energy-dependent metal efflux systems involving ATP ases and chemiosmotic ion/proton pumps are reported for As, Cr and Cd resistance in many bacteria (Roane and Pepper, 2000).

Advantage of the bioremediation:

1. The action of microbes or other biological systems to degrade environmental pollutants (Dua *et al.,* 2002).

2. It can be applied in situ without the removal and transport of polluted soil and without the disturbance of the soil matrix (Kuiper *et al.,* 2004).

3. The bacterial degradation usually results in complete mineralization of the pollutant (Heitzer and Sayler, 1993).

Mechanisms of Biosorption:

A number of methods have been developed for the removal of heavy metals from liquid wastes such as evaporation, precipitation, ion exchange, electroplanting, membrane process and etc. Each methods has its own merits and demerits. Biosorption is a process, which represents a biotechnological innovation as well as a cost effective excellent tool for removing heavy metals from aqueous solutions. This process involves solid phase (sorbent, biological material) as well as liquid phase (solvent) containing a dissolved species to be sorbed (sorbate, metal ions). The degree of sorbent affinity for the sorbate determines its distribution between the solid and liquid phases. Advantages of biosorption are low cost, possibility of metal recovery, high efficiency, regeneration of biosorbent, minimization of chemical and biological sludge while disadvantage is no potential for biologically altering the metal valency state (Ahluwalia and Goyal, 2007).

The mechanism of metal biosorption is very complicated process by living cells has two steps process. In the first step, metal ions are adsorbed to the surface of cells by interactions between metals and functional groups displayed on the surface of cell. All the metal ions before gaining access to the cell membrane and cell cytoplasm come across the cell wall which consist of a variety of polysaccharides and proteins and hence offers a number of active sites capable of binding metal ions. Difference in the cell wall composition among the different groups of microorganism viz., algae, bacteria, fungi and cyanobacteria cause significant differences in the type and amount of metal iron binding to them (Das *et al.*, 2008). In passive biosorption is metabolism independent and proceeds rapidly by any one or a combination of the following metal binding mechanisms: coordination, ion exchange, physical adsorption (eq. electrostatic), complexation and inorganic micro precipitation. In passive biosorption is a dynamic equilibrium of reversible adsorption-desorption. Metal ions bound on the surface can be eluted by other ions, chelating agents or acids. In second step, due to active biosorption, metal ions penetrate the cell membrane and enter into the cell. The major factor affects the biosorption processes are pH, temperature (Aksu *et al.*, 1992), initial metal ion concentration (Galun *et al.*, 1987; Friis and Keith, 1998), biomass concentration in solution (Fourest and Roux, 1992).

Bioaugmentation: It is a method to improve degradation and enhance the transformation rate of xenobioticsby the injection (seeding) of specific microbes, able to degrade the xenobiotics of interest. Examples of microbes which has genetic potential to mineralize recalcitrant pollutants such as PAHs, chlorinated aliphatics and aromatics, nitroaromatics, and long-chain alkanes (Heidelberg *et al.*, 2002). These microbes can be can be genetically modified strains equipped with catabolicplasmids, containing the relevant degradation genes, but also can be wild-type isolates(Yee *et al.*, 1998). Dean-Ross (1987) and Heitkamp and Cerniglia (1989) reported that analysis of the efficiency of microbial degradation showed that chlorinated derivatives,especially, are difficult to metabolize and frequently are degraded by means of co-metabolism. Here, the oxidation of non-growth substrates during the growth of an organism on another carbon or energy source is known as co-metabolism.

Mechanisms of metal-microbe interactions: The study of the interactions between microorganisms and metals may be helpful to understand the relations of toxic metals with higher organisms such as mammals and plants. Devi and Joshi (2012) reported that to understanding and exploring the microbe-metal interaction have resulted in an upsurge in the research interest with their importance in various high-through put biotechnological applications such as biofuels cell, biosensor and most promisingly in microbe mediated nanomaterial synthesis. Mandal *et al.* (2006) reported that identification of the microbial ligands/ cellular processes involved in metal sequestration has lead to the development of engineered organisms with various cell surface displays that facilitate their applications in industrial catalysis, biosorption, bioremediation, biosensor technology and biofuel. Moreover, genetic engineering can be used to improve the bioremediation capacity of both plants and microbes and further improve their capacity in bioremediation of heavy metals and radio nuclide (Joshi *et al.*, 2014).

Mechanisms of Bioaccumulation:

Bioaccumulation of contaminants in aquatic organisms is commonly described by a simple mass transfer kinetic model and it occurs via several routes, including absorption across the cuticle, respiratory surfaces and assimilation from ingested media. Bioaccumulation mainly focus on its potential for microbes or other biological cells to accumulate heavy metal species from the ambient environment. Bioaccumulation is a process to adsorb and to accumulate metal species by biological substrate, such as agricultural waste, plant residues and microorganisms, is a promising approach for selective binding of metals and metalloids (Yang *et al.*, 2015). It also have great potential for adorption and preconcentration of ultra-trace levels of heavy metals for their analysis and speciation. Bioaccumulation include both the fraction of chemical that enters the organism by absorption of freely dissolved contaminant and that assimilated from ingestion of contaminated media (Suedel *et al.*, 1994). It can be understood as a simple partitioning process in which contaminants accumulate in living organisms in proportion to their lipophilicity. However, there are a variety of biological and physical factors that can influence bioaccumulation and result in deviations from the expected steady-state distribution of contaminants (Fisher, 1995). Rand and Petrocelli (1985) reported that once a chemical enters an aquatic food chain, it is possible for the contaminant to be passed from one trophic level to the next via the process known as trophic transfer. Aquatic systems in which the bulk of contaminants are retained in sediment, slower transfer process, such as bioaccumulation and tropics transfer, may predominate (Landrum and Robbins, 1990). Fish are considered very sensitive indicators of heavy metal contamination in aquatic ecosystems, as they are vertebrates whose life cycle is completely aquatic. Although the bioaccumulation of heavy metals in fish has been well known (Pritchard, 1993). Carriquiriborde and Ronco (2008) examined on bioaccumulation of three heavy metals, Cd^{2+}, Cu^{2+} and Cr^{6+} in liver and gill of juvenile specimens of Pejerrey (*Odontesthes bonariensis*). Here, the cumulative kinetics of the three metals was separately analyzed for each tissue by assuming first-order kinetics and passive diffusion mechanisms of bioaccumulation. Blanco *et al.* (2014) results indicated that bioaccumulation of Cd^{2+}, Cu^{2+} and Cr^{6+} is described by a combination of a concentration-independent and saturable uptake kineticsin both organs with a unidirectional path of elimination from gills to liver to waterborne. Finally, the good agreement between the parameter values predicted by the model and previously published data suggests that our modelling approach may shed light on the mechanisms of heavy metal bioaccumulationin other species. The rate of metal bioaccumulation linearly depends on concentrations (as with simple diffusion), saturable kinetics, or the combination of both, should be considered. Moreover, the kinetics of bioaccumulation of a heavy metal in a given tissue/organ could be non-independent of the transport and concentration of the same metal in another tissue/organ.

Bacteria can sequester heavy metals and radio nuclides through passive sorption mechanisms involving charged constituents on the cells surface (biosorption) or by an energy consuming process involving the transport to the interior of the cell (bioaccumulation). The current understanding of sorption and accumulation of heavy metals and radio nuclides by bacteria attached to mineral surfaces compromises our ability to predict microbiological immobilization of contaminants in subsurface and sedimentary environments. Techniques are needed to obtain non intrusive, in situ measurements of metal speciation, mobile, soluble, low molecular weight and colloidal organic phases that bind metals and radio nuclides. New experimental approaches need to be developed that incorporate processes of bioaccumulation and biosorption in studies of iron and sulfate reduction and biooxidations (Gorby and Geesey, 2000).

Mechanisms of Biomineralization:

Biomineralization is the process by which inorganic substance are produced on organic template within an organism. On the other word, process by which organisms from minerals. For instance, formation of sea urchin spines, shells and formation of bone and teeth within the human body. Another example of biomineralization is the pearl layer of the Akoya pearl oyster, in which single crystals of aragonite, a calcium carbonate, are packed in an orderly fashion within a microscope space. The proteins that fill in the surrounding gaps contribute to the creation of a beautiful sheen. Similarly, teeth are assemblies of single apatite crystals, and enamel in particular is a high hardness tissue comprised of densely packed fine apatite with an extremely high degree of crystallinity (Onuma *et al.*, 2012). Although biomineralization beneficially forms tissues necessary for maintaining normal organism function, it also produces substances that adversely affect organism health. Example: gout, kidney stones, urinary stone, gallstones, vein wall deposit and cholesterol accumulation in humans.

Mechanisms of biotransformation:

Biotransformation process is an old as civilization itself, as evidence of wine making from 5400 B.C. has been found in Hajji FiruzTepe, Iran. Fernandes *et al.* (2003) reported that microorganisms are widely used in the industrial-scale production of steroids, with the products being used in a wide variety of medicinal situations, including inflammation, transplantation, adrenal insufficiency, osteoporosis, coronary heart disease, HIV and cancer. There are several researchers has focus due to owing to historical reason, bacteria and fungi have usually been the microorganisms of choice in the biotransformation of exogeneous compounds (Mahato and Banerjee, 1985; Ahmad *et al.*, 1991; Mahato and Majumdar, 1993; Fernandes *et al.*, 2003; Faramarzi *et al.*, 2008).

Mechanisms of bioleaching

Bioleaching is define as the dissolution of metals from their mineral source by certain naturally occurring microogamisms. On the other way, the use of microorganisms to transfer elements so that the elements can extracted from a material when water is filtered through it (Atlas and Bartha, 1992). Hansford and Miller (1993) called as biooxidation. In the case of copper, copper sulfide is microbially oxidized to copper sulfate and metal values are present in the aqueous phase. In gold mining operations, biooxidation is used as a pretreatment process to (partly) remove pyrite or arsenopyrite. This process is also known as biobeneficiation where solid materials are refined and unwanted impurities are removed (Groudev, 1999). Mandal *et al.* (1996) reported that mobilization of elements from solid materials mediated by bacteria and fungi then it called as 'biomining', 'bioextraction' or 'biorecovery'. Biohydrometallurgy represents an inter disciplinary field where aspects of microbiology (especially geomicrobiology), geochemistry,biotechnology, hydrometallurgy, mineralogy, geology, chemical engineering, and mining engineering are combined.

Factors Influencing Bioleaching: For physicochemical parameters of a bioleaching environment affect on light, pressure, surface tension, pH, temperature, water potential, redox potential, oxygen content and availability, carbon dioxide content, nutrient availability, mass transfer, iron(III) concentration and presence of inhibitors. For microbiological parameters of a bioleaching environment depends on adaptation abilities of microorganisms, microbial activities, microbial diversity, population density and metal tolerance. For properties of the minerals to be leached depends on mineral composition, mineral type, mineral dissemination, surface area, porosity, grain size, hydrophobicity, galvanic interactions and formation of secondary minerals. Similarly, for processing depends on leaching mode (in situ, heap, dump, or tank leaching), pulp density stirring rate (in case of tank leaching operations) and heap geometry (in case of heap leaching).

The following equations describe the "direct"and "indirect" mechanism for the oxidation of pyrite (Sand *et al.,* 1995). However, the model of 'direct' and 'indirect' metal leaching is still under discussion.

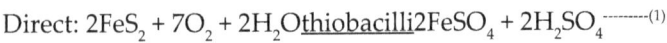

Direct: $2FeS_2 + 7O_2 + 2H_2O \underline{thiobacilli} 2FeSO_4 + 2H_2SO_4$ --------(1)

Indirect:

$4FeSO_4 + O_2 + 2H_2SO_4 \underline{Thiobacillus ferrooxidans, L. ferrooxidans} 2Fe(SO_4)_3 + 2H_2O-$
$$---(ii)$$

$FeS2 + Fe2(SO4)3 \underline{Chemical\ oxidation} 3FeSO4 + 2S$ -----(iii)

$2S + 3O_2 + H_2O \underline{T.\ thiooxidans} 2H_2SO_4$ -----------(iv)

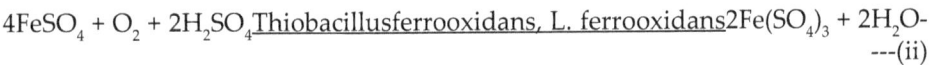

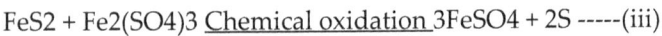

Role of Biofilms in Bioremediation

'Biofilm' was proposed as an operational term to refer to organisms and organic material that accumulate at mineral surfaces or other solid phases. Costerton *et al.* (1987) defines that biofilm as an assemblage of microorganisms comprising of microbial species attached to a biological or inert surface and encased in a self-synthesized matrix comprising of water, proteins, carbohydrates and extracellular DNA. Biofilm forming bacteria are adapted to survive and suited for bioremediation as they compete with nutrients and oxygen and observations of tolerance of biofilms towards harsh environment found way in the process of bioremediation. Jobson *et al.* (1974) and Amadi (1992) reported that organic waste materials such as poultry litter (PL) and Coir pith (CP) to the soil facilitates aeration through small pores and increases the water holding capacity of the soil, thus enhancing bioremediation. Bacterial biofilms exist within indigenous populations near the heavily contaminated sites to better persist, survive and manage the harsh environment. Expressions of genes vary within the biofilms and are distinctive relative to free floating planktonic cells. Differential gene expressions within biofilms are owing to variable local concentration of nutrients and oxygen within biofilm matrix and division of labour among microbes. Such variable gene expression may be important for degradation of varied pollutants by numerous metabolic pathways. An important consideration for biofilm formation in microorganisms is chemotaxis and flagellar dependent motility (Pratt and Kolter, 1999). Responses such as swimming, swarming, twitching motility, chemotaxis, quorum sensing in presence of xenobiotics commonly present in soil and water assist microbes to coordinate movement towards pollutant and improved biodegradation (Lacal *et al.*, 2013). Biofilms are interact with nutrients and adsorb pollutants from the environmental monitor in aqueous bodies, develops rapidly and offer easy sampling methods (Fuchs *et al.*, 1997; Peacock *et al.*, 2004). The multi properties of biofilm may be assessed for monitoring environmental pollution including change in biomass, species composition, pigment production, photosynthesis and enzymatic activity. Species present in river biofilms varies seasonally and on the level of pollution (Brummer *et al.*, 2000). Arini *et al.* (2012) reported that heavy environmental contamination such as Zinc and Cadmium can influence the species diversity within biofilms and as such species diversity estimation of microbial biofilms can also indicate environmental pollution. Rather than convention techniques, there are several modern molecular biology methods such as ribosomal spacer analysis, denaturing gradient gel electrophoresis and terminal restriction fragment length polymorphism has been used to estimate species diversity within biofilms (Dorigo *et al.*, 2002; Kostanjsek *et al.*, 2005; Szabo *et al.*, 2008; Bricheux *et al.*, 2013).

Biostimulationis the widely accepted bioremediation procedure in which indigenous microorganisms by the addition of nutrients, as input of large quantities of carbon sources tends to result in rapid depletion of the available pools of major inorganic nutrients, such as nitrogen, phosphorus and addition to increase microbial metabolism and to encourage bioremediation.

Hazardous agricultural compound and its remediation techniques

Fungicides, herbicides, insecticides, rodenticides, nematicides, molluscicides, algaecides and plant growth regulators comes under agricultural hazardous compounds. Most of these compounds are organochlorinated (cause cancer, neurological disorder and teratogenic effects), organophosphate (affect nervous system and reproductive system) or carbamate (symptoms such as sweating, lacrimation, hypersalivation and convulsion by inhibiting the enzyme activity of acetylcholinesterase) (Suzuki and Watanabe, 2005; Vaccari *et al.*, 2006; Yair *et al.*, 2008). Jeyaratnam (1985) reported that dissemination of these hazardous compounds into the food stuff has resulted in serious health implications. Apart from the traditional techniques for removing contaminants from soil is decreasing.

Additionally, conventional methods such as land-filling, recycling, excavation, incineration, stabilization, and vitrification have enormous pitfall. These techniques are expensive, adverse effects on the environment and even kills the native flora and fauna. The most important drawback is the production of toxic and hazardous intermediate compounds which affect entire ecosystem (Chaudhary, 2016).

Polycyclic aromatic hydrocarbons (PAHs): It is widespread organic micro pollutants, as fossil fuels continue to be used as a principal source of energy, which are resistant to environmental degradation due to their highly hydrophobic nature (depending upon the number of fused rings) and adverse effect on ecosystem such as toxic, carcinogenic, mutagenic, teratogenic, and resistant to biodegradation (Dreyer *et al.,* 2005; Fernández-Luqueño *et al.,* 2008; Gan *et al.,* 2009; Sayara *et al.,* 2011; Nduka *et al.,* 2012). The main input sources of PAH pollutions are leaches from old storage tanks, road surfaces, and domestic waste; oil spills; tanker leakage; incomplete fossil fuel combustion; and seepage from natural oil reservoirs (Morganand Watkinson, 1989).Once these pollutants enter the soil, they are trapped into soil pores and immobilized by adsorption to the soil matrix (Reddy and Sethunathan, 1983; Aprill and Sims, 1990). Eom *et al.* (2007) and Okere and Semple (2012) reported that soil is a major reservoir of PAHs among other environmental media and stores more than 90 per cent of the total PAHs found in the environment, because of the sorbing capacity of soil particles. The causes of PAHs mainly in environment due to refinery and coal-gasification waste and chemical present in flue gas condensates, which spread through transportation, disposal, and accidental spills of petroleum products, making them important global environmental pollutants (Eom *et al.,* 2007; Nduka *et al.,* 2012; Okere and Semple, 2012). The specifically technologies for PAHs include solvent extraction, bioremediation, phytoremediation, (Kumar, 2018) chemical oxidation, photocatalytic degradation, electrokinetic remediation, thermal treatment and integrated remediation technologies (Gan *et al.,* 2009). Due to their ubiquitous occurrence, recalcitrance, bioaccumulation potential and carcinogenic activity, the PAHs have gathered significant environmental concern. A number of bacterial species are known to degrade PAHs and most of them are isolated from contaminated soil or sediments. Haemophilus spp., Mycobacterium spp., Pseudomonas aeruginosa, Pseudomons fluoresens, Paenibacillus spp., Rhodococcus spp. are some of the commonly studied PAH-degrading bacteria (Haritash and Kaushik, 2009). PAHs are oxygenase, dehydrogenase and lignolytic enzymes. The biodegradation of PAHs has been observed under both aerobic and anaerobic conditions and the rate can be enhanced by physical/chemical pretreatment of contaminated soil (Gan *et al.,* 2009; Haritash and Kaushik, 2009). Addition of biosurfactant-producing bacteria and light oils can increase the bioavailability of PAHs and metabolic potential of the bacterial community. The supplementation of contaminated soils with compost materials can also enhance biodegradation without long-term accumulation of extractable polar and more available intermediates. Aquatic weeds Typha spp. and Scirpus lacustris have been used in horizontal-vertical macrophyte based wetlands to treat PAHs. An integrated approach of physical, chemical, and biological degradation may be adopted to get synergistically enhanced removal rates and to treat/remediate the contaminated sites in an ecologically favorable process (Haritash and Kaushik, 2009).

References:

Ahluwalia, S.S. and Goyal, D. (2007). Microbial and plant derived biomass for removal of heavy metals from waste water. *Bioresour. Technol.* 98: 2243-2257.

Ahmad, S., Roy, P.K., Khan, A.W., Basu, S.K. and Johri, B.N. (1991). Microbial transformation of sterols to C19 steroids by Rhodoccusequi. *World J Microbiol Biotechnol.* 7: 557-561.

Aksu, Z., Sag, Y. and Keith, M. (1992). The biosorption of copper by C. vulgaris and Z. ramigera. *Environ. Technol.* 13: 579-586.

Amadi, A.J. (1992). A double control approach to assessing the effect of remediation of preplan ting oil pollution on maize growth. *Delta Agric. J. Nigeria,* 1: 1-6.

Aprill, W. and Sims, R.C. (1990). Evaluation of the use of prairie grasses for stimulating polycyclic aromatic hydrocarbon treatment in soil. *Chemosphere,* 20: 253-265.

Arini, A., Feurtet-Mazel, A., Maury-Brachet, R., Pokrovsky, O., Coste, M. and Delmas, F. (2012). Recovery potential of periphytic biofilms translocated in artificial streams after industrial contamination (Cd and Zn). *Ecotoxicology,* 21: 1403-1414.

Atlas, R.M. and Bartha, R. (1992). Microbial ecology: fundamentals and applications. The Benjamin Cummings, Redwood City, CA.

Balaji, V., Arulazhagan, P. and Ebenezer, P. (2014). Enzymatic bioremediation of polyaromatic hydrocarbons by fungal consortia enriched from petroleum contaminated soil and oil seeds. *J. Environ. Biol.* 35: 521-529.

Bassey, D.E. and Grigson, S.J.W. (2011). Degradation of Benzyldimethyl Hexadecylammonium Chloride by Bacillus Niabensis and Thalassospira Sp. Isolated From Marine Sediments, Toxicol. *Environ. Chem.* 93(1): 44-56.

Blanco, M.V., Cattoni, D.I., Carriquriborde, P., Grigera, J.R. and Chara, O. (2014). Kinetics of bioaccumulation of heavy metals in*Odontesthesbonariensis* is explained by a single and common mechanism. *Ecological Modeling,* 274: 50-56.

Bricheux, G., Le Moal, G., Hennequin, C., Coffe, G., Donnadieu, F., Portelli, C., Bohatier, I. and Forestier, C. (2013). Characterization and evolution of natural aquatic biofilm communities exposed in vitro to herbicides. *Ecotoxicol. Environ. Saf.* 88: 126-134.

Brummer, I.H., Fehr, W. and Wagner-Dobler, I. (2000). Biofilm community structure in polluted rivers: abundance of dominant phylogenetic groups over a complete annual cycle. *Appl. Environ. Microbiol.* 66: 3078-3082.

Carriquiriborde, P. and Ronco, A.E. (2008). Distinctive accumulation patterns of

Cd(II) Cu(II), and Cr(VI) in tissue of the South American teleost, pejerrey (*Odontesthesbonariensis*). *Aquatic Toxicology*. 86: 313-322.

Chaudhary, D.K. (2016). Bioremediation: An Eco-friendly Approach for Polluted Agricultural Soil. *Emer. Life Sci. Res.* 2(1): 73-75.

Chaudhary, D.K. and Kim, J. (2016). Novosphingobiumnaphthae sp. nov., from oil contaminated soil. Int. J. Syst. Evol. Microbiol.doi: 10.1099/ijsem.0.001164.

Costerton, J.W., Cheng, K.J., Geesey, G.G., Ladd, T.I., Nickel, J.C., Dasgupta, M. and Marrie, T.J. (1987). Bacterial biofilms in nature and disease. Annu. Rev. Microbiol. 41: 435-464.

Das, N., Vimala, R. and Karthika, P. (2008). Biosorption of heavy metals-an overview. *Indian Journal of Biotechnology*. 7: 159-169.

Dean-Ross, D. (1987). Biodegradation of toxic wastes in soil. *ASM News,* 53: 490-492.

Devi, L.S. and Joshi, S.R. (2012). Antimicrobial and synergistic effects of silver nanoparticles synthesized using soil fungi of high altitudes of Eastern Himalaya. *Mycobiology*. 40: 27-34.

Dhankhar, R. and Hooda, A. (2011). Fungal biosorptionean alternative to meet the challenges of heavy metal pollution in aqueous solutions. *Environ. Technol.* 32: 467-491.

Dorigo, U., Bérard, A. and Humbert, J.F. (2002). Comparison of Eukaryotic Phytobenthic Community Composition in a Polluted River by Partial 18S rRNA Gene Cloning and Sequencing. *Microbial Ecology,* 44: 372-380.

Dreyer, A., Radke, M., Turunen, J. and Blodau, C. (2005). Long-term change of polycyclic aromatic hydrocarbon deposition to peatlands of eastern Canada. *Environ. Sci. Technol*. 39: 3918-3924.

Dua, M., Sethunathan, N., and Johri, A. K. (2002). Biotechnology and bioremediation: successes and limitations. *Appl. Microbiol. Biotechnol*. 59:143-152.

Eom, I.C., Rast, C., Veber, A.M. and Vasseur, P. (2007). Ecotoxicity of a polycyclic aromatic hydrocarbon (PAH)-contaminated soil. *Ecotoxicol. Environ. Saf.* 67: 190-205.

Erguven, G.O., Bayhan, H., Ikizoglu, B., Kanat, G. and Nuhoglu, Y. (2016). The capacity of some newly bacteria and fungi for biodegradation of herbicide trifluralin under agiatedculture media. *Cell. Mol. Biol.* 62(6): 74-79.

Faramarzi, M.A., Adrangi, S. and Yazdi, M.T. (2008). Microalgal biotransformation of Steroids. *Journal of Phycology*. 44(1): 27-37.

Fernandes, P., Cruz, A., Angelova, B., Pinheiro, H.M. and Cabral, J.M.S. (2003). Microbial conversion of steroid compounds: recent developments. *Enzyme Microbial Technol*. 32: 688-705.

Fernández-Luqueño, F., Marsch, R., Espinosa-Victoria, D., Thalasso, F., Hidalgo

Lara, M.E., Munive. A., Luna-Guido, M.L. and Dendooven, L. (2008). Remediation of PAHs in a saline-alkaline soil amended with wastewater sludge and the effect on dynamics of C and N. *Sci. Total Environ.* 402: 18-28.

Fisher, S.W. (1995). Mechanisms of bioaccumulation in aquatic systems. Rev. Environ. *ContamToxicol.* 142: 87-117.

Fonti, V., Beolchini, F., Rocchetti, L. and Dell'Anno, A. (2015). Bioremediation of contaminated marine sediments can enhance metal mobility due to changes of bacterial diversity. *Water Res.* 68: 637-650.

Fourest, E. and Roux, J.C. (1992). Heavy metal biosorption by fungal mycelial by products: Mechanisms and I Influence of pH.*Appl. Microbiol. Biotechnol.* 37: 399-403.

Friis, M. and Keith, M. (1998). Biosorption of uranium and lead by Streptomyces longwodensis. *Biotechnol. Bioeng.* 35: 320-325.

Fuchs, S., Haritopoulou, T., Schäfer, M. and Wilhelmi, M. (1997). Heavy metals in freshwater ecosystems introduced by urban rainwater runoff-Monitoring of suspended solids, river sediments and biofilms. *Water Sci. Technol.* 36: 277-282.

Galun, M., Galun, E., Siegel, B.Z., keller, P., Lehr, H.(1987). Removal of metal ions from aqueous solutions by Penicillum biomass: Kinetic and uptake parameters. *Water, Air, Soil Pollut.* 33: 359-371.

Gan, S., Lau, E.V. and Ng, H.K. (2009). Remediation of soils contaminated with polycyclic aromatic hydrocarbons (PAHs). *Journal of Hazardous Materials,* 172(2-3): 532-549.

Gaur, N., Flora, G., Yadav, M. and Tiwari, A. (2014). A review with recent advancements on bioremediation-based abolition of heavy metals. *Environ Sci. Process Impacts,* 16: 180-193.

Gorby, Y.A. and Geesey, G. (2000). Metal/ Microbe interactions workshop. Sponsered by Natural and ccelerated bioremediation research program office of energy research. U.S. Department of Energy. October 11-13, 2000.

Groudev, S.N. (1999). Microbial detoxification of heavy metals in soil. *MineraliaSlovaca.* 28: 335-338.

GWRTAC(1997). Remediation of metals-contaminated soils and groundwater. Tech. Rep. TE-97-01, GWRTAC, Pittsburgh, Pa, USA, 1997, GWRTAC-E Series.

Hansford, G.S. and Miller, D.M. (1993). Biooxidation of a gold-bearing pyrite-arseoopyrite concentrate. *FEMS Microbiology Reviews,* 11(3): 175-181.

Haritash, A.K. and Kaushik, C.P. (2009). Biodegradation aspects of Polycyclic Aromatic Hydrocarbons (PAHs): A review. *Journal of Hazardous Materials,* 169(1-3): 1-15.

Heidelberg, J.F., Paulsen, I.T., Nelson, K.E., Gaidos, E.J., Nelson, W.C., Read, T.D., and Eisen, J.A. (2002). Genome sequence of the dissimilatory metal ion-reducing bacterium *Shewanellaoneidensis*. *Nature Biotechnol*. 1:1-6.

Heitkamp, M.A. and Cerniglia, C.E. (1989). Polycyclic aromatic hydrocarbon degrading by a mycobacterium sp. in microcosms containing sediment and water from a pristine ecosystem. *Appl. Environ. Microbiol*. 55: 1968-1973.

Heitzer, A. and Sayler, G.S. (1993). Monitoring the efficacy of bioremediation. *Tibtech*. 11: 334-343.

Jeyaratnam, J. (1985). Health problems of pesticide usage in the third world. *Br. J. Ind. Med*. 42: 505-506.

Jobson, A. M., Mclaughlin, M., Cook, F. D.,Westlake, D.W.S. (1974). Effects of amendment on microbial utilization of oil applied to soil.*Appl. Microbiol*.27, 166-171.

Joshi, S.R., Kalita, D., Kumar, R., Nongkhlaw, M. and Swer, P.B. (2014). *Metal-microbe interaction and bioremediation*. In: Gupa, D.K. and Walther, C. (eds.), Radionuclide Contamination and Remediation through plants. Springer International Publishing, Switzerland, Pp-235-251.

Kamaludeen, S.P., Arunkumar, K.R., Avudainayagam, S. and Ramasamy, K. (2003). Bioremediation of chromium contaminated environments. *Indian J. Exp. Biol*. 41: 972-985.

Khosravi, F., Savaghebi, G.H. and Farah, B.H. (2009). Effect of potassium chloride onCd uptake by colza in a polluted soil. *Water Soil J*. 23: 28-35.

Kostanjsek, R., Lapanje, A., Drobne, D., Perovic, S., Perovic, A., Zidar, P., Strus, I, Hollert, H. and Karama, G. (2005). Bacterial community structure analyses to assess pollution of water and sediments in the Lake Shkodra/Skadar, Balkan Peninsula. *Environ Sci. Pollut. Res. Int*. 12: 361-368.

Kuiper, I., Lagendijk, E.L., Bloemberg, G.V. and Lugtenberg, Ben J.J. (2004). Rhizoremediation: A beneficial plant-microbe interaction. MPMI. 17(1): 6-15.

Kumar, V. (2018). Phyloremediation as a Potentially, Promising Technology: Prospects and future. In: Gautam, A. and PAthak, C. (Eds). Metallic coulamination and its Toxicity. Daya Publishing House (Astral) New Delhi, India PP:207-229.

Lacal, J., Reyes-Darias, J.A., García-Fontana, C., Ramos, J.L. and Krell, T. (2013). Tactic responses to pollutants and their potential to increase biodegradation efficiency. *J. Appl. Microbiol*. 114: 923-933.

Landrum, P.F. and Robbins, J.A. (1990). *Bioavailability of sediment associated contaminants: A review and simulation model*. In: Baudo, R., Geisy, J.P., Muntau, H. (eds), Sediments: Chemistry and Toxicity of In-Place Pollutants. Lewis Publishers, Chelsea, MI. pp 237–263.

Lim, S.J. (2015). Biodegradation: Enzymes Evolution. *Journal of Bioremediation & Biodegradation*, 6(6): e168.

Ling, W., Shen, Q., Gao, Y., Gu, X. and Yang, Z. (2007). Use of bentonite to control the release of copper from contaminated soils. *Australian Journal of Soil Research*, 45(8): 618-623.

Mahato, S.B. and Banerjee, S. (1985). Steroid transformations by microorganisms - III.*Phytochemistry*, 24(7): 1403-1421.

Mahato, S.B. and Majumdar, I. (1993). Current trends in microbial biotransformation. *Phytochemistry*, 34(4): 883-898.

Mandal, B.K., Chowdhury, T.R., Samanta, G., Basu, G.K., Chowdhury, P.P., Chanda, C.R., Lodh, D., Karan, N.K., Dhar, R.K., Tamili, D.K., Das, S., Saha, K.C. and Chakraborti. D. (1996). Arsenic in groundwater in seven districts of West Bengal, India- the biggest arsenic calamity in the world. *Current Science*, 70(2): 976-986.

Mandal, D., Bolander, M.E., Mukhopadhyay, D., Sarkar, G. and Mukherjee, P. (2006). The use of microorganisms for the formation of metal nanoparticles and their application. *Appl. Microbiol. Biotechnol.* 69: 485-492.

Mani, D. and Kumar, C. (2013). Biotechnological advances in bioremediation of heavy metals contaminated ecosystems: an overview with special reference to phytoremediation. *Int. J. Environ. Sci. Technol.* 11: 843-872.

McGuinness, M. and Dowling, D. (2009). Plant-Associated Bacterial Degradation of Toxic Organic Compounds in Soil. *Int. J. Environ. Res. Public Health,* 6: 2226-2247.

McLaughlin, M.J., Hamon, R.E., McLaren, R.G., Speir, T.W. and Rogers, S.L. (2000b). Review: a bioavailability-based rationale for controlling metal and metalloid contamination of agricultural land in Australia and New Zealand. *Australian Journal of Soil Research*, 38(6): 1037-1086.

McLaughlin, M.J., Zarcinas, B.A., Stevens, D.P. and Cook, N. (2000a). Soil testing for heavy metals.*Communications in Soil Science and Plant Analysis*, 31(11-14): 1661-1700.

Merugu, R., Prashanthi, Y., Sarojini, T. and Badgu, N. (2014). Bioremediation of waste waters by the anoxygenic photosynthetic bacterium *Rhodobactersphaeroides* SMR 009. *Int. J. Res. Environ. Sci. Technol.* 4: 16-19.

Mishra, A. and Malik, A. (2014). Novel fungal consortium for bioremediation of metals and dyes from mixed waste stream. *Bioresour. Technol.* 171: 217-226.

Mitra, A. and Mukhopadhyay, S. (2016). Biofilm mediated decontamination of pollutants from the environment. *AIMS Bioengineering*, 3(1): 44-59.

Mohsenzade, F. Chehregani, A. and Akbari, M. (2012). Evaluation of oil removal efficiency and enzymatic activity in some fungal strains for bioremediation of petroleum-polluted soils. Iran *J. Environ. Health Eng.* 9: 26-34.

Morel, J.L., Echevarria, G. and Goncharova, N. (2002). Phytoremediation of Metal-Contaminated Soils. IOS Press, Amsterdam, and Springer in conjunction with the NATO Public Diplomacy Division, pp: 1-11.

Morgan, P., and Watkinson, R.J. (1989). Hydrocarbon degradation in soil and methods for soil biotreatment. *Crit. Rev. Biotechnol.* 8: 305-333.

Nduka, J.K., Umeh, L.N., Okerulu, I.O., Umedum, L.N. and Okoye, H.N. (2012). Utilization of Different Microbes in Bioremediation of Hydrocarbon Contaminated Soils Stimulated With Inorganic and Organic Fertilizers. *J. Pet. Environ. Biotechnol.* 03: 116.

Okere, U.V. and Semple, K.T. (2012). Biodegradation of PAHs in pristine soils from different climatic regions. *J. Bioremediation Biodegrad.* 01: 1-11.

Onuma, K., Tsuji, T. and Lijima, M. (2012). Biomineralization: Mechanisms of hydroxyapatite crystal growth. In: Liu, X.Y. (eds), Bioinspiration from Nano to Micro Scales. Biological and Medical Physics, *Bimedical Engineering*. Springer. Pp-135-159.

Peacock, A.D., Chang, Y.J., Istok, J.D., Krumholz, L., Geyer, R., Kinsall, B., Watson, D., Sublette, K.L., White, D.C. (2004). Utilization of microbial biofilms as monitors of bioremediation. *Microb. Ecol.* 47: 284-292.

Picado, A., Nogueira, A., Baeta-Hall, L., Mendonça, E., de Fátima Rodrigues, M., do CéuSàágua, M., Martins, A. and Anselmo, A.M. (2001). Land farming in a PAH-contaminated soil. *J. Environ. Sci. Heal.* A36: 1579-1588.

Pratt, L.A. and Kolter, R. (1999). Genetic analyses of bacterial biofilm formation. *Curr. Opin. Microbiol.* 2: 598-603.

Pritchard, J.B. (1993).Aquatic toxicology: past, present, and prospects. Environmental Health Perspectives, 100.

Rand, G.M. and Peirocelli, S.R. (1985). Fundamentals of Aquatic Toxicology: Methods and Applications. Hemisphere Publishing Co., New York, p 652.

Raymond, J., Rogers, T., Shonnard, D. and Kline, A. (2001). A review of structure-based biodegradation estimation methods. *J. Hazard. Mater.* 84: 189-215.

Reddy, B.R. and Sethunathan, N. (1983). Mineralization of parathion in the rice rhizosphere. *Appl. Environ. Microbiol.* 45: 826-829.

Roane, T.M. and Pepper, I.L. (2000). Microorganisms and metal pollution. In Environmental Microbiology; Maier, I.L., Pepper, C.B., Eds., Gerba, Academic Press: London, UK, p. 55.

Sand, W., Gehrke, T., Hallmann, R. and Schippers, A. (1995). Sulfur chemistry, biofilm, and the (in)direct attack mechanism-a critical evaluation of bacterial leaching. *Appl. Microbiol. Biotechnol.* 43: 961-966.

Sayara, T., Borràs, E., Caminal, G., Sarrà, M. and Sánchez, A. (2011). Bioremediation of PAHs-contaminated soil through composting: influence of bioaugmentation and biostimulation on contaminant biodegradation. *Int. Biodeterior Biodegrad.* 65: 859-865.

Sikkema, J., de Bont, J.A. and Poolman, B. (1995). Mechanisms of membrane toxicity of hydrocarbons. *Microbiol. Rev.* 59: 201-222.

Suedel, B.C., Boraczek, J.A., Peddicord, R.K., Clifford, P.A. and Dillon, T.M. (1994).

Trophic transfer and biomagnification potential of contaminants in aquatic ecosystems. *Rev. Environ. ContamToxicol.* 136: 21–84.

Suzuki, O. and Watanabe, K. (2005). Drugs and Poisons in Humans - A Handbook of Practical Analysis. In: Carbamate Pesticides. 559-570.

Szabo, K.E., Makk, J., Kiss, K.T., Eiler, A., Acs, E., Toth, B., Kiss, K.A. and Bertilsson, S. (2008). Sequential colonization by river periphytonanalysed by microscopy and molecular fingerprinting. *Freshwater Biology*, 53: 1359-1371.

Tang, C.Y., Criddle, Q.S., Fu, C.S. and Leckie, J.O. (2007). Effect of flux (transmembrane pressure) and membranes properties on fouling and rejection of reverse osmosis and nanofiltration membranes treating perfluorooctane sulfonate containing waste water. *Jou. Enviro. Sci. Tech.* 41: 2008-2014.

Uqab, B., Mudasir, S., Qayoom, A. and Nazir, C. (2016). Bioremediation: A Management Tool. *Journal of Bioremediation & Biodegradation*, 7: 331.

Vaccari, D.A., Strom, P.F. and Alleman, J.E. (2006). Environmental Biology for Engineers and Scientists. John Wiley & Sons.

Verma, P., George, K., Singh, S., Juwarkar, A. and Singh, R. (2006). Modeling rhizofiltration: heavy metal uptake by plant roots. *Environmental Modeling and Assessment*, 11: 387-394.

Xiezhi, Y., Jieming, C. and Ming, H.M. (2005). Earthworm mycorrhiza interaction on Cduptake and growth of ryegrass. *Soil Biol. Biochem.* 37: 195-201.

Yair, S, Ofer, B., Arik, E., Shai, S., Yossi, R., Tzvika, D. and Amir, K. (2008). Organophosphate Degrading Microorganisms and Enzymes as Biocatalysts in Environmental and Personal Decontamination. Applications. *Crit. Rev. Biotechnol.* 28: 265-275.

Yang, T., Chen, M.L. and Wang, J.H. (2015). Genetic and chemical modification of cells for selective separation and analysis of heavy metals of biological or environmental significance. *Trends in Analytical Chemistry*, 66: 90-102.

Yee, D.C., Maynard, J.A. and Wood, T.K. (1998). Rhizoremediation of trichloroethylene by a recombinant, root-colonizing Pseudomonas fluorescens strain expressing toluene ortho- monooxygenase constitutively. *Appl. Environ. Microbiol.* 64: 112-118.

Role of Earthworms in Managing Soil Contamination

Payal Garg and Geetanjali Kaushik

Centre for Rural Development and Technology, Indian Institute of Technology, New Delhi-110016

Abstract

Rapidly growing Indian cities generate huge quantities of solid wastes on a daily basis. On account of enormous environmental implications the management of this organic solid waste has become a major problem. Heavy metal contamination due to dumping of contaminated solid wastes has been reported. Application of vermicomposting on large scale for urban as well as rural waste disposal has been highlighted by research. This chapter discusses the types of earthworms and their role in maintaining soil fertility, vermicomposting, soil erosion control and land reclamation.

Keywords: Wastes, management, pollution control, vermicomposting, land reclamation

Earthworms and pollution control

Management of organic solid waste has become a major problem due to environmental implications, thus attracting the attention of researchers. Many publications related to solid waste generation, problems and disposal practices have emerged throughout the world to generate environmental awareness. The urgency for recycling and composting of solid wastes is felt as countless dumping areas in and around urban and rural settlements, causes environmental hazards (Chaturvedi, 1994). The long-term consequences of organic wastes dumping near the agricultural fields in Hyderabad have been reported (Rao and Shantaram, 1994). It is found that soil upto 30-cm depth was contaminated with heavy metals due to dumping of contaminated solid wastes. Unprecedented increase in the cost of chemical fertilizers hit the small and marginal farmers badly. An obvious way to mitigate their difficulty is generation of fertilizers at the village level through recycling of wastes. Bhiday (1994) asserted that the feasibility of using vermiculture on large scale for urban as well as rural waste disposal is very high and it is a right kind of technology for recycling such wastes.

Earthworms encourage growth of microorganisms in their gut providing ideal conditions there in (Bhawalkar, 1993). Werner and Cuvas (1996) reported that

Cuba has more than 170 vermicomposting centers that are engaged in producing earthworm casting for use as fertilizers in tobacco farming. Mani (1997) suggested that vermicomposting of animal and agricultural waste results in higher nutrient content within six weeks. Singh and Rai (1997) asserted that earthworm farming and vermicomposting is a boon for sustainable agriculture in India. Status report on solid waste management by vermicomposting in India has been prepared by Vasudevan *et al.* (2001). Santra and Bhowmik (2001) studied that vermicomposting is attaining a special significance for the abatement of pollution hazards created by large amount of organic wastes in our country and also reduce the demand for chemical fertilizers.

Earthworms for healthy soil

Earthworms soil's intimate friends and benefactor, has since long been helping soil in respiration, nutrition, excretion and various other vital activities. Through its characteristic functions of breaking, grinding, churning, assimilation and tunneling, earthworm has proved to be soils mouth, stomach and intestine. The earthworm is nature's marvelous control machine as it eats practically almost any matter except rubber, plastic, glass and metal. Alvares (1984) reported that 20 million worms could handle 80 metric tones of pulp sludge daily or about five grams of sludge per worm per day and whatever worm eats get transformed into organic fertilizer. The worm's castings are rich in nitrates, phosphates and potash in other words a rich source of organic fertilizer.

Earthworms, the major secondary decomposer macro-fauna have a very major role in hasting up of the rate of decomposition and also in improving the structural properties of soil. They serve as agent for pollution control and for amelioration of the soil. The lumbricid earthworms are dominantly distributed in the temperate soil as megascolecid earthworms are in the tropical and subtropical soils. Roles of earthworms in soil fertility are very high basically due to life activities of earthworms in soil and their importance in utilization of earthworms for human welfare viz. Vermiculturing and Vermicomposting. Earthworms establish well in lands receiving organic manure. Role of earthworms in the breakdown of organic debris on soil surface and in soil turn over process was first highlighted by Darwin (1881).The involvement of earthworms in the composting process decreases the time of stabilization of the waste and produces an efficient organic pool with energy reserves as vermicompost. The sludge from both agro based industries and domestic sewage plants can be a food source for composting earthworms with suitable organic amendments such as plant litter or animal waste. Earthworms in nature promote infiltration of water in ground as they make soil porous and promote drainage. Earthworms increase natural fertility of soil. They would help in decreasing the chemical fertilizer use and increase the fertilizer efficiency in biological and natural way. Chemical fertilizers are well known to affect soil chemistry, deplenish soil micro-nutrients and cause water pollution.

The role of earthworms in breaking down of dead plants and animal residues in soil was first studied by Charles Darwin in 1881. Later on many of the research workers studied the mechanisms of conversion of organic matter into humus by introducing vermiculture in the field. Earthworms can be defined as invertebrates belonging to phylem annelida, order oligochaeta, class clitellata, which live in soil. From the days of ancient greek philosophers and Darwin to the present day, these lowly creatures have been recognized as master builders of top soil and man's fellow tillers. Aristotle has referred to earthworms as 'the intestine of the earth'. Darwin (1881) remarked,

"The plough is one of the most ancient and most valuable man's inventions, but long before it existed, the land was In fact regularly ploughed and still continues to be thus ploughed by earthworms".

Types of earthworms

More than 4200 species of Oligochaetas are known in the word. Of these, 280 microderill and remaining about 3920 belongs to megadrilli (earthworms). In the Indian sub-continent earthworms also form bulk of the Oligochaete fauna. They are representing by 509 species and 67 genera, indicating a high degree of diversity in this region as compared to other areas. Earthworms form a major component of the soil biota and they together with a large number of other organisms constitute the soil community. The chief source of food to the soil biota is the litter contributed by plants. Although the dead plant tissues constitute the bulk of the food ingested by the earthworms, living microorganisms, fungi, microfauna and mesofauna and their dead tissue are also ingested as an important part of the diet (Parle, 1963, Piearce, 1978). Though earthworms are generally called as saprophages, they can be classified based on the feeding habits (Lee, 1985) into detrivores and geophages.

Detrivores feed at or near the soil surface, mainly on plant litter or dead roots and other plant debris in the organic matter rich surface soil horizon or on mammalian dung. These

Worms are classified as humus formers and comprise of epigeic and anecic forms. *Perinyx excavatus, Eisenia fetida, Eudrillus eugineae, Lampito mauritti, Polypheretima elongate, Octochaetona serrata* and *Octochaetona surensis* are a few examples of detrivores earthworms.

Geophages worms feed deeper beneath the surface, ingesting large quantities of organically rich soil. The worms are generally called as humus feeders and comprise of the endogeic worms. *Octochaetona thrustoni* is one such earthworm commonly available in Madras.

A different classification has been proposed by Bouche (1977) laying stress on ecological strategies. He classified earthworms into epigeics, anecics and endogeics. The epigeics have no effect on the soil structure as they generally can not dig. They are efficient agents of combination and fragmentation of leaf litter. These are broadly classified as phytophagous earthworms. The anecic feed on the leaf litter mixed with the soil of the upper horizons. They may also produce surface casts. These are called as geophytophagous earthworms.

Ecological categories of earthworms

Based on the nature and their vertical distribution in soil, earthworms can be distinguished into three different ecological categories (Bouche, 1977),

i) **Epigeic:** In nature epigeic worms live in the top soil, and duff layer on the soil surface. These small, deeply pigmented worms have a poor burrowing ability, preferring instead an environment of loose organic litter or loose topsoil rich in organic matter to deeper soils. Epigeic species feed in organic surface debris and have adapted beautifully to the rapidly shifting, dynamic environment of the soil surface. They are tolerant to some disturbance, moderate to high rate of cocoons production and short life cycle.

ii) **Endogeic:** Endogeic worms build complex lateral burrow systems through all layers of the upper mineral soil. These worms rarely come to the surface; instead spending their lives in these burrow systems where they feed on decayed organic matter and bits of mineral soil. They are the only category of worm which actually eats significant volumes of soil and not strictly the organic component. Endogeic worms tend to be medium sized, tolerant to some disturbance, moderate to high rate of cocoon production, life-cycle not very long and pale colored.

iii) **Anecic:** Anecic worms (like the common night crawler Lumbricus terrestris) build permanent, vertical burrows that extend from the soil surface down through the upper mineral soil layer. These worm species coat their burrows with mucous which stabilizes the burrow so it does not collapse, and build little mounds (called middens) of stone and castings outside the burrow opening. They also tend to be very large worms and have bellies with less pigmentation than their backs. These worms intolerant to disturbance, low rate of cocoons production and long life cycle (Julka, 1986). It is reported that mostly epigeic and some endogeic species are suitable for vermicomposting.

Suitable earthworm species for vermicomposting

Species which are identified as potentially useful species to break down organic wastes include *E.fetida, Dendrobaena venta* and *Lumbricus rubellus*, from temperate areas and *Eudrilus eugeiae* and *perionyx excavatus* from the tropics. In different parts of India degradation of organic waste was done successfully using a number of species viz., *E.fetida, Aporrectodea caliginosa, Eudrilus eugeniae, Perionyx arboricola* etc (Goswami and Kalita, 2000). *E.fetida* (Savigny) is abundant worldwide for vermicomposting and a large literature is available on its performance in converting organic waste into quality vermicompost and its high growth rate on various organic substrates (Kaviraj and Sharma, 2003; Kaushik and Garg, 2003; Thimmaiah, 2001). It is an epigeic earthworm species which lives in organic wastes and require high moisture content, adequate amount of suitable organic material and dark conditions for proper growth and development (Chaudhari and Bhattacharjee, 2002; Gundai and Edwards, 2003). *E.fetida* proved to be a suitable species for vermicomposting application, because of its rapid growth rate, reproductive potentials and occurrence in rich organic substrates in nature (Neuhauser *et al.*, 1980).

Influence of earthworm activities on soil properties

Earthworm activities are direct action of feeding and burrowing along with related biological activities. Earthworm eats its way through soil and organic humus etc.

Earthworms improve the soil fertility in following ways:

Influence on soil pH: The pH of the intestinal contents of earthworms is remarkably stable around neutral to slightly alkaline. This can have a profound effect on the overall level of soil pH and on the course of organic decomposition. In neutral or slightly alkaline conditions bacterial activity is favored, leading to more complete breakdown of organic compounds, and a multi-type humus.

Physical decomposition: The passage of organic material through the earthworm gut results in the physical decomposition due to the muscular grinding action of gizzard, aided by ingestion of silica granules. This provides considerably enhanced surface area for microbial decomposition.

Humus formation: The process of humus formation is often characterized by the selective breakdown of cellulose. The end product is a complex mixture of various organic acids, amino acids, polyphenols and sugars such as glucose, galactose, mannose, arabinose and xylose. Lignin fibers are present in raw humus and peat but are degraded to polyphenols in well-decomposed humus.

Improvement of soil structure: The physical communication of organic particles, the amelioration of soil pH, the enhancement of microbial decomposition activity-all these results of earthworm activity contribute to soil fertility. Burrowing of earthworms brings about tillage of soil upto 3 meters without adversely affecting plants in any manner excepting some species in special situations. This also accompanies breakdown of soil particles and mixing of soil nutrients and bacteria in digestive process as well as with deposit of casts of various levels. These affect automatic conversion of organic wastes. These micronized soil particles lead increase of particle surface area which leads increased moisture absorption and holding and air circulation etc. This also increases microbial action. Casts is fine biofertilizer having upto 1000 times more microbes than in surrounding soil. Increased porosity eventually increase percolation of water generally referred as charging of sub-soil water. This in turn, led maintenance of soil temperature which adds to toleration in soil faunal and floral components. These reduce severity of soil fluctuations essential for plant growth.

Soil enrichment: Application of vermicompost significantly improved the physical properties of all the soil types under study. Earthworms physically mix the contents of the deeper layers and make the soils loose and porous. Their body exudates improve the water holding capacity of soil and promote establishment of microorganisms (Kale, 1994). Rani and Srivastava (1997) conducted an experiment on rice with full dose of nitrogen replaced by one-third and quarter of N as vermicompost. Compared with N fertilizer alone, vermicompost application showed increase in grain yield and yield components of rice. It also improves the availability of Phosphorous and potassium as well as micro-nutrients.

Role of earthworms in soil erosion control

Soil aggregates are formed by the adhesion of mineral and organic particles. The shape and physical packing of which influence the aeration, infiltration of water, water holding capacity and surface area etc. Formation of aggregates makes the soil well aerated and drained. These aggregates are mineral granules joined together in such a way that they can resist wetting, erosion or compaction and remain loose when the soil is dry or wet. Most of the workers agree that earthworm casts contain more stable aggregates than the surrounding soil. In an experiment, the percentage of aggregates in soil to which earthworms are added was compared with that in soil without earthworms. In general, earthworm burrows and structural aggregation due to their casting activities promote water entry into the soil and therefore reduce surface runoff. In addition, polysaccharide gums produced by the bacteria by the soil, as the soil passes through the gut of the earthworm (Dash, 1978). The production of the polysaccharides gums is enhanced by the presence of the organic components in the ingested material.

Another possibility of the cause of stability is that the soil particles are cemented by calcium humate which is derived from the interaction of the ingested organic matter and calcite excreted by the earthworm's calciferous glands.

Nutrient Availability from Vermicompost

About 5-10 percent of the ingested material is absorbed into the tissue of earthworms for their growth and metabolic activity and rest is excreted as cast or vermicompost. Vermicomposting are a highly enriched kind of biofertilizer. It is more chemically neutral than the surrounding soil. The nutrient levels in the vermicompost depend on the nature of the organic waste used as food source for earthworms .It is found that a heterogeneous waste mix will have balanced level of nutrients than from any one particular waste. Vermicompost contains most nutrients in plant available forms such as nitrates, phosphates, exchangeable calcium, soluble potassium etc (Edward, 1998) and large surface area that provide many micro sites for microbial activity and for the strong retention of nutrients. The nutrient status of vermicompost is given in table 1 (Kale, 1995). The vermicompost is considered an excellent product since it is homogenous, has reduced level of contaminants and tends to hold more nutrients over a longer period without impacting the environment (Ndegwa and Thompson, 2000). Earthworm cast typically have high N content which suggests that they would be good sources of plant N (Parmelee and crossley, 1988; Ruz-Jerez *et al.*, 1992). In addition to increased N availability, C, P, K, Ca and Mg availability in the casts is also greater than initial feed material (Daniel and Anderson, 1992; Lavelle and Martin, 1992; Basker *et al.*, 1993 Orozco *et al.*, 1996).

Table 1: Nutrient content of vermicompost

Nutrient	Percentage
Organic carbon	9.15-17.98
Total nitrogen	0.5-1.5
Available Phosphorous	0.1-0.3
Available Potassium	0.15-0.56
Available Sodium (ppm)	0.06-0.03
Copper	2.0-9.5*
Iron	2.0-9.3*
Zinc	5.7-11.5*
Available sulphur	128-548*

*Source: Kale, 1995, *Values in ppm*

The beneficial effect of earthworm cast has been observed in both horticultural plants (Saciragic and Dzelilovic, 1986; Hidalgo, 1999) and in agronomic crops (Pashanasi *et al.*, 1996). More than the regular macro-nutrients, vermicompost contributes to the supply of micro-nutrients essential for crops. The stimulatory effect of vermicompost for nutrient uptake, growth and yield of crops is linked to the secretion of earthworms and the associated microbes mixed with the cast.

Enzymatic activities in earthworm's cast and vermicompost

Enzyme activities have been postulated as indicators of the decomposition process (Diaz-Burgos *et al.*, 1992; Garcia *et al.*, 1993). Vermicompost contains enzymes such as proteases, amylases, lipases, cellulases and chitinases; which continue to disintegrate organic matter even after earthworms have been excreted and hence, vermicompost is believed to have additional attributes of providing enzymes and hormones which stimulate plant growth (Abbasi and Ramasamy, 1999; Atiyeh *et al.*, 2001; Chaoui *et al.*, 2003). In several studies, higher enzyme activities were measured in worm casts than in surface soil. Earthworm casts have been shown to have enhanced carbohydrase, protease (Ross and Cairns, 1982), phosphatase and dehydrogenase activities than the surrounding un-ingested soil (Tiwari *et al.*, 1989; Mulongoy and Bedoret; 1989). Ranganathan and Vinotha (1998) showed enhanced enzyme activity in the gut of *E.eugeniae* when it was fed with pressmud.

Plant Growth Hormones and Regulatores in Vermicompost

Vermicompost contain plant growth regulators and other plant growth influencing materials (Tomati *et al.*, 1988; Grappelli *et al.*, 1987; Atiyeh *et al.*, 2002). The beneficial influence of worm cast has been related to the biological factors like gibberellin, cytokinins and auxins released due to microbial activity of the microbes harboured in the cast (Brown, 1995). It is reported that certain metabolites produced by earthworms may be responsible for stimulating plant growth. It is considered that earthworms release into the soil certain vitamins and similar substances which may be B group vitamins or some pro vitamin D or free amino acids. Vermicompost also contain large amount of humic substances (Senesi *et al.*,1992; Masciandaro *et al.*, 1997) and some of the effects of these substances on plant growth have been shown to be very similar to the effects of soil-applied plant growth regulators or hormones (Muscolo *et al.*, 1999).The antibacterial activity of coelomic fluid of earthworms is accounted and this activity is only directed against the highly pathogenic soil bacteria. Thus it could be deducted that earthworms apart from encouraging the establishment of beneficial micro-organisms to some extent can also inhibit the soil borne pathogens.

Some of the organic acids that are isolated from the body fluid of earthworms and their cast have similar response as that of plant growth promoter substances. The nutrient value of worm castings is not high compared to chemical fertilizers. The key factor is microbial activity. Microbial activity is 10-20 times higher than in the soil and organic matter that the worm ingests. The most important effect of earthworms may be the stimulation of microbial activity that occurs in casts. This enhances the transformation of soluble nitrogen into microbial protein; preventing their loss by leaching to the lower horizons of the soil.

Vermicompost is better than chemical fertilizers in economical and ecological aspects. Replacing costly yet deadly chemicals with cheap yet friendly vermicompost will ensure sustainable food production.

Use of vermicompost as manure has multifold benefits. They are:

  i)   Healthy soil with soil organisms;

  ii) Limited external inputs;

  iii) Cost effective farming practices and healthy food;

  iv) Problems of leaching and mineralization of nutrients are reduced.

Use of Vermicompost for Crop Production

Vermicompost is one such material which is being experimented as a growth media and carried for biofertilizer. The use of vermicompost as soil amendments can have many positive effects on soil physical characteristics following high rates of application. It is rich in nitrogen, phosphorous, potassium, carbon and organic matter all of which are essential for the growth of microbes. High levels of organic humic matter soil amendment in the form of compost improve soil structure by increasing porosity and reducing the bulk density of an amended soil. Polysaccharides and other polymeric substances present in organic matter act as aggregating compounds (Masciandaro *et al.*, 2000) and increase micropores in the soil. The improvements to soil physical structure, soil fertility, and soil microbiological properties associated with compost application all promote plant growth, as a growth medium for transplants and a soil amendment for field crops. Several studies have evaluated the effect of vermicompost-amended potting media on plant growth greenhouse production. Generally, potting medium with 10 to 20% vermicompost by volume provides adequate fertilization for transplant growth (Subler *et al.*, 1998; Atiyeh *et al.*, 2000a; Ozores-Hampton and Vavrina (2002). In one study, germination rates of greenhouse tomatoes increased up to 15% when vermicomposted pig manure was mixed with potting medium at 20, 30, and 40% by volume. The highest marketable yield of fruit was reported in the 20% mixture. Treatments consisting of 100% vermicompost led to smaller growth and fewer leaves than other treatments, due to high-moisture and possible phytotoxicity (Atiyeh *et al.*, 2000a). Sagar *et al.* (2002) compared growth of *Ocimum sanctum*, a crop used for essential oil, in vermicomposted cow manure, farmyard manure, urea, and a control growth medium. While oil yield was 15 g kg–1 in all treatments, leaf weight was 25% higher in vermicompost treatments than control and plant weight 52% higher. Ranganathan and Cristopher (1994) reported that the vermicompost not only helped to protect fertility of the soil but also boosted productivity depending on crops, season and other factors and also enhanced the quality of end products.

Use of Earthworms in Vermiwash Production- A Liquid Manure

Foliar Sprays are used as a part of agronomic practices for crop production; vermiwash is one such new concept in this context. Worm worked soils have burrows formed by the earthworms. Bacteria richly inhabit these burrows, also called as the drilospheres. Water passing through these passages washes the nutrients from these burrows to the roots to be absorbed by the plants. This principle is applied in the preparation of vermiwash. It is a collection of excretory products and mucus secretions of earthworms along with nutrients from the soil organic molecules. If it collected properly is a clear and transparent, pale yellow coloured fluid. The utility of vermiwash as a biocide, either singly or when mixed with botanical pesticides is still under investigation. Following methods of vermiwash preparations are described by few scientists:

First Method

Ismail (1997) has suggested a method of preparation of vermiwash using a barrel in which the layers of cattle

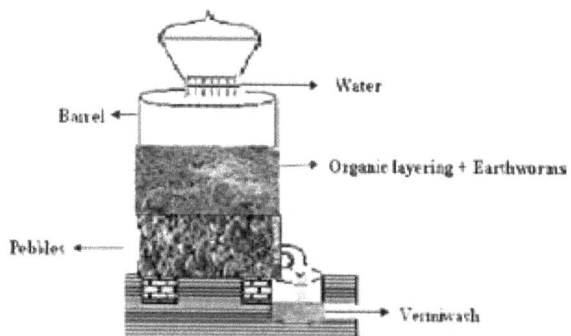

Fig.1: Method of Vermiwash Preparation

dung and hay were placed on the top of the layer of soil (Fig 1). The epigeic earthworms were introduced to produce compost at faster rate and anecics were put to produce a large number of drilospheres. The vermiwash was collected from the bottom of the barrel after sprinkling water through perforated mud from the above. It was suggested to use the vermiwash for foliar spray either as such or dilution with water or 10%cow's urine.

The physico-chemical characteristics of vermiwash are as follows:

Parameter	Vermiwash
pH	6.9
Chlorides (ppm)	110.00
Sulfates (ppm)	177.00
Inorganic phosphate (μg/l)	50.9
Amonical nitrogen (ppm)	below detectable level
Potassium (ppm)	69.00
Sodium (ppm)	122.00
Total hardness (ppm)	375.00
Calcium hardness (ppm)	175.00
Magnesium hardness (ppm)	200.00
BOD (Biological oxygen demand) (ppm)	4.60
COD (Chemical oxygen demand) (ppm)	97.00

Source; Ismail, 1997

Second Method

A method described by Kale (1998) consisted of an outer and an inner vessels. The inner vessel would have an outlet at the lower side of the vessel. The inner vessel was filled with decomposing organic matter and about 1 to 2 kg earthworms were accommodated in 12 to 16 litre capacity vessel. (Fig 2).As the earthworms were accommodated in waste, water was slowly added into vessel in excess. The excess water flowing out through the outlet as thick syrupy fluid was collected in the outer vessel. The fluid so collected was siphoned out and after diluting was used as foliar spray to different crops. In another method, the cement tank was built at an elevated place from which they want to collect the wash. The slope provided in

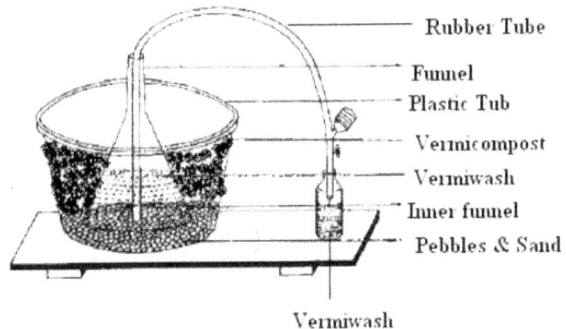

Fig. 2: Method of vermiwash preparation

the tank provided scope for excess water to flow out in drops as thick syrupy emulsion through a small outlet. This was collected in a container and stored in bottles. Before using, it was diluted and sprayed to crops.

Third method

Karuna *et al.* (1999) also prepared earthworm exudates and used successfully for foliar spray to Anthuriums. In this method, clitellated earthworms of *Eudrillus eugeniae* were collected from the culture bins. These worms were cultured on organic waste comprising mostly of leaf litter, weeds, stubble mixed with one part of cow dung by volume in the form of slurry. Earthworms weighing one kg were placed into a dry enamel tray for 15 to 20 min to clear out the cast that will be excreted due to handling. Earthworms were then carefully separated from their excreta and then added into glass plastic bowl having 500 ml of distilled water having temperature of 37°C-40°C (luke warm) (fig.3). The worms were agitated for 3 to 5 min. and removed and added into another bowl containing 500 ml of water at room temperature (25°C-27°C) to rinse them thoroughly to collect the exudates adhering to its body wall before releasing back to the culture bins. The contents in the two bowls were mixed to use as a spray. The exudates thus collected were a syrupy, light yellow fluid.

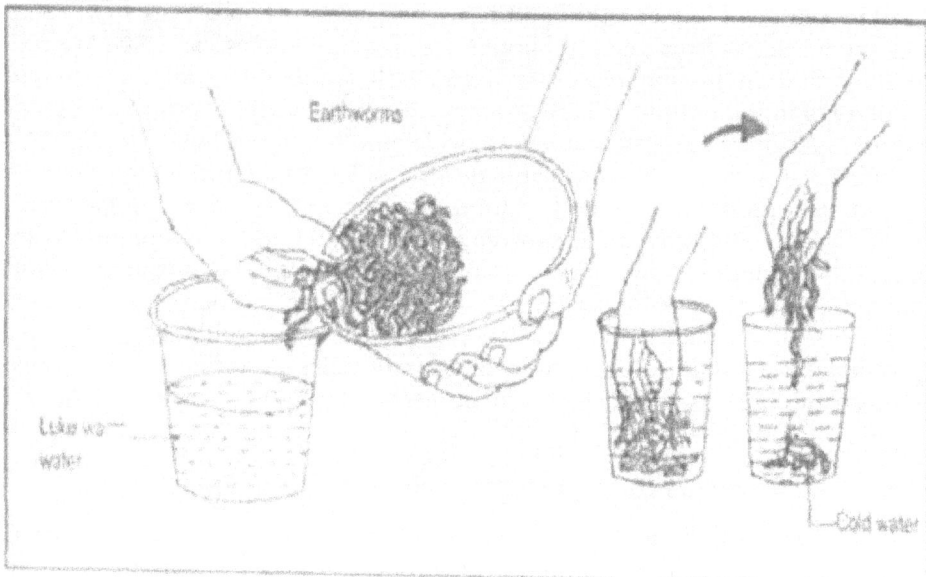

Fig.3: Method of vermiwash preparation

Fourth method

Giraddi (2001) evaluated a method of extraction of earthworm wash, a plant promoter substance. In order to have simple devices of collection of earthworm secretions an equipment using GI sheets or plastic crates. The equipment consists of culturing compartment, joined to collection chamber at the bottom. While, water sieve is kept enclosed on the culturing chamber. Both water sieve and culturing compartment have got small holes to allow water from sieve to culturing compartment to collection chamber is regulated by inserting a divider in between to chambers. Number of earthworms, quantity and quality of raw material used, duration of earthworm culturing and time allowed for water to stagnate in the culturing compartment would decide the quality of Vermiwash to be used.

Vermiwash have enzymes, secretions of earthworms which would stimulate the growth and yield of crops and even develop resistance in crops receiving this spray. Such a preparation would certainly have the soluble plant nutrients apart from some organic acids and mucus of earthworms and microbes (Shivsubramanian and Ganeshkumar, 2004). But so far there are no experimental evidences to quantify the effect of such spray. Chemical composition of vermiwash varies with the types of substrates used for vermiwash (Lourduraj and Yaday, 2008).

Effect of Vermiwash on Yield and Quality of Crops

As in crops, to tackle the pest problem, indiscriminate use of synthetic chemical insecticides was undertaken. This had led to many serious problems like environmental contamination by way of pesticide residues, development of resistance in pests to pesticides, pest resurgence, destruction of natural enemies etc. There is a need for developing pest management strategies, which are eco-friendly, and environmentally safe. Vermiwash seems to possess an inherent property of in soil fertility, is very important in maintaining balance in acting not only as a fertilizer but also as a mild biocide an ecosystem (Shuster *et al.*, 2000). The fresh vermiwash harbours a large number of beneficial microorganisms that help in plant growth and protects it from a number of infestations. Ismail (1997) reported that vermiwash can be sprayed on plants as a foliar spray for improving quality and yields of Okra crop. Foliar application of vermiwash prepared by both the methods resulted into considerable increase in total nitrogen, phosphorous and potassium uptake by seasonal chrysanthemum, marigold and china aster in comparison to NPK alone gave indication of quick absorption of the above nutrients through foliage for better nourishment of these flowering plants (Todkari and Talashilkar, 2001).

Use of earthworms in land improvement and reclamation

It is now well established that the introduction of earthworms into soils that have no earthworms or have only low natural earthworm population is usually beneficial in terms of plant growth and crop yield. Stock dill and Cossens (1966) have successfully introduced earthworms into pastures in New Zealand that lacked native earthworms. Following establishment of the earthworm populations, the organic matter at the base of the grass disappeared and compacted soils with poor structure were transformed into deep friable top soils. Some of the methods of earthworm inoculation for the improvement and reclamation of land tried by various workers are described below:

Direct release of worms

A large number of deep working species of earthworms are introduced by sufficiently moistening and liming of acidic soils. If food is a limiting factor, addition of organic material such as municipal sludge or animal wastes allow their earlier establishment to facilitate stabilization of temperature and moisture conditions. The mature worms should be placed at the rate of $150/m^2$. Addition of earthworms to soil seems particularly promising in reclaiming flooded areas that are subsequently drained and put into cultivation with tree crops. In such cases, earthworms can be introduced at the rate of 100-180 worms per tree in the basis around the base of tree and mulched with cattle dung and leaf litter.

As soil Blocks

Earthworms can be transferred to new localities by moving blocks of soil cut from the surface at localities where target species are common. Such Blocks of soil are commonly introduced into small areas and supplied with organic matter as food.

As Turf Pieces

In this method, turf pieces with earthworms are cut and placed on the ground surface at 10 m spacing between them. To cut turf pieces, sod machines are developed in New Zealand. This method is useful for introduction of earthworms into pasture devoid of earthworm before.

As Vermicomposting

Surface application of vermicompost to degrade soil can improve its structure and fertility. Such lands may be planted with grass cover for the establishment of earthworms.

Barley and Keleing (1964) successfully introduced *Aporrectodea caliginosa* and the megascolecid species, *Microscolex dubius* into newly sown, irrigated pasture on sandy loam soil in Australia with corresponding significant improvement is soil structure, loss of organic materials and increased productivity. Ghilarov and Mamajev (1966) inoculated earthworms into reclaimed irrigated land in Uzbekistan and reported considerable improvement in soil structure and fertility. Van Rhee (1969 and 1971) introduced earthworms into polders that had been drained and reclaimed from the sea in the Netherlands and reported that this accelerated the development of normal soil structure and facilitated the return of the polders to productive use. He found that grass yields of these polders increased upto four times and clover yields upto ten times, after inoculation with earthworms. The dry matter yield of spring wheat was doubled by inoculation with earthworms and fruit trees were established much more readily than in uninoculated soil.

Conclusion

With rapidly growing Indian cities becoming powerhouses of commercial and industrial activity huge quantities of wastes are expected to be generated every day. Indiscriminate disposal of organic wastes is associated with various forms of environmental hazards. In this background earthworm through application of vemicomposting offers a ray of hope for waste management and pollution control.

References

Abbasi S.A., Ramasamy E.V. 1999. Anaerobic digestion of high solid waste. In: Proceedings of Eighth National symposium on Environment IGCAR, Kalpakkam, India, 20-22 July, pp220-224.

Alvares, C. (1984). Agriculture and Ecology, Science for Villages, 79: 4-5.

Atiyeh R.M., Arancon N.Q., Edwards C.A., Metzger J.D., 2001. The influence of earthworm-processed pig manure on the growth and productivity of marigolds. Bioresour. Technol.81, 103–108.

Atiyeh R.M., Lee S., Edwards C.A., Arancon N.Q., Metzger J.D., 2002. The influence of humic acids derived from earthworms-processed organic wastes on plant growth. Bioresour. Technol. 84, 7–14.

Atiyeh R.M., Dominguez J., Subler S., Edwards C.A., 2000. Changes in biochemical properties of cow manure processed by earthworms (*Eisenia andrei*) and their effects on plant-growth. Pedobiologia 44, 709–724.

Barley K.P., Kleining C.R. 1964. The occupation of newly irrigated land by earthworms. Australian Journal of Science.26:290.

Basker A, Macgregor A.N., Kirman J.H. 1993. Exchangeable potassium and other cations in non ingested soil and casts of two species of pasture earthworms. Soil Biol. Biochem. 25, 1673-1677.

Bhawalkar U.S., Bhawalkar V.U., 1993. Vermiculture biotechnology. In:Thampan, P.K. (Ed.), Organics in soil health and crop production. Peekay tree crops development foundation, Cochin, pp. 69–85.

Bhiday M.R. 1994. Earthworms in Agriculture. Indian Farming 43 (12):31-34.

Bouche MB.1977. Strategies lombriciennes. Ecol.Bull. (stocholm).25, 122-132.

Brown G.G., 1995. How do earthworms affect micro-floral and faunal community diversity? Plant and Soil 170, 209–231.

Chaoui H.I., Zibilske L.M. Ohno T., 2003. Effects of earthworm casts and compost on soil microbial activity and plant nutrient availability. *Soil Biology and Biochemistry*, 35, 295–302.

Chaturvedi A.S. (1994). The chain of recycling. In: The Hindustan Times. Aug 14.

Chaudhuri P.S. and Bhattacharjee G. (2002). Cocoon production, morphology, hatching pattern and fecundity in seven tropical earthworm species- a laboratory based investigation. *J.Biosci.*, 27(3),283-294.

Daniel, O. and Anderson, J.M. (1992). Microbial biomass and activity in contrasting soil materials after passage through the gut of the earthworm *Lumbricus rubellus* Hoffmeister. *Soil Biol. Biochem.*, 24, 465–470.

Darwin C. (1881). The formation of vegetable mould through the action of worms with observation of their habits. Murray, London 326 pp.

Dash M.C. (1978). Role of earthworms in the decomposer system. Off print from Glimpses of Ecology (Prof. R. Misra Commemoration Vol.) Publs Int. Scientific Publn. 309-406.

Diaz-Burgos M.A., Ceccanti B. and Polo, A. (1992). Monitoring biochemical activity during sewage sludge composting. *Biol. Fert. Soil*, 16, 145-150.

Garcia C., Hernandez T., Costa F., Ceccanti C., Masciandaro G. and Ciardi C. (1993). A study of biochemical parameters of composted and fresh municipal wastes. *Biores. Technol,.* 44, 17-23.

Ghilarow M.S. and Mamajev B.M. (1966).Uberdie ansteduling von regenwurmem in den artesisch bewasserten oasen der Wurste Kyst-Kum. *Pedobiologia*, 6: 197-218.

Giraddi, R.S. (2001). Method of extraction of earthworm wash, A plant promoter substance, Souvenir and Abstracts. A paper presented in Silver Jubilee Celebrations of the Indian Society of Soil Biology and ecology and VII National symposium on soil Biology and Ecology held at U.A.S, Bangalore during Nov.7-9, 2001:pp 66.

Goswami, B. and Kalita, M.C. (2000). Efficiency of some indigenous earthworms species of Assam and its characterization through vermitechnology. *Indian J. Environ. Ecoplan.*, 3: 351–354.

Grappelli A., Galli E. and Tomati, U. (1987). Earthworm casting effect on *Agaricus bisporus* fructification. *Agrochimica*, 21: 457–462

Gundai, B. and Edwards, C. A. (2003). The effect of multiple application of different organic wastes on the growth, fecundity and survival of *Eisenia fetida* (savigny) (Lumbricidae). *Pedobiol.*, 47 (4), 321-330.

Hidalgo, P. (1999). Earthworm castings increase germination rate and seedling development of cucumber. Mississippi Agricultural and forestry Experiment Station, Research Report 22, 6.

Ismail, S.A. (1997). Vermicology – the biology of earthworms. Orient Longman Limited, Hyderabad (A.P.), India.

Julka, J.M. (1986). Earthworms resources of India Proc. Nat. Sem. Org. waste utilization, Vermicomp. part B: verms and Vermicomposting. Dash, R.C., Senapathi, B.K. and Mishra, P.C. (eds.);pp1-7.

Kale, R.D. (1998). Earthworms: nature's gift for utilization of organic wastes. In: Edwards, C.A. (Ed.), Earthworm Ecology. Soil and Water Conservation Society. Ankeny, Lowa St. Lucie Press, New York, pp. 355–373.

Kale, R.(1995). Vermicomposting has a bright scope. *Indian Silk*, 34:6-9.

Kale, R.D. (1994). Agro-waste composting through earthworms. Proc. National meeting on waste recycling. Centre of Scince for village, Wardha.

Karuna, K., Patil, C.R., Narayanswamy and Kale, R.D. (1999). Stimulatory effect of earthworm body fluid (Vermiwash) on Crinkle Red Variety of _Anthurium andreanum_ Lind. _Crop Res.,_ 17 (2): 253-257.

Kaushik P. and Garg, V.K. (2003). Vermicomposting of mixed solid textile mill sludge and cow dung with the epigeic earthworm _Eisenia foetida. Bioresour. Technol.,_ 90: 311–316.

Lavelle P., Martin A., 1992. Small-scale and large-scale effects of endogeic earthworms on soil organic matter dynamics in soils of the humic-tropics. Soil Biol. Biochem. 12, 149–1498.

Lee K.E. 1985. Earthworms, their ecology and relationships with soil and land use. Academic Press, Sydney, Australia, pp. 188–194.

Lourduraj A.C. and Yadav B.K. (2007).Effect of organic manures applied to rice crop on microbial population and enzyme activity in post harvest soil. J.Ecobiol ,20(2), 139-44.

Mani D (1997): Organic farming and biofertilizer for sustainable agriculture. Yojna. 41 (9): 26.

Masciandaro, G., Ceccanti, B. and Garcia, C. (1997). Soil agro-ecological management: fertirrigation and vermicompost treatments. _Bioresource Technology,_ 59: 199–206.

Masciandaro, G., Ceccanti, B., Ronchi, V. and Bauer, C. (2000). Kinetic parameters of dehydrogenase and inorganic fertilizers. _Biol. Fertil. Soils,_ 32 (6): 579-587.

Mulongoy K. and Bedoret, A. (1989). Properties of worm casts and surface soils under various plant covers in the humid tropics. _Soil Biol. Biochem.,_ 21: 197–203.

Muscolo, A., Bovalo, F., Gionfriddo, F. and Nardi, S. (1999). Earthworm humic matter produces auxin-like effects on _Daucus carota_ cell growth and nitrate metabolism. _Soil Biol. Biochem.,_ 31, 1303–1311.

Ndegwa P.M., Thompson S.A., Das K.C. (2000). Effects of stocking density and feeding rate on vermicomposting of biosolids. _Biores. Technol.,_ 71, 5–12.

Neuhauser, E.F., Hartenstein, R. and Kaplan, D.L. (1980). Growth of the earthworm _Eisenia foetida_ in relation to population density and food rationing. _OIKOS,_ 35, 93–98.

Orozco, F.H., Cegarra, J., Trujillo, L.M. and Roig, A. (1996). Vermicomposting of coffee pulp using the earthworm _Eisenia foetida_: effects on C and N contents and the availability of nutrients. _Biol. Fertil. Soil.,_ 22: 162–166.

Ozores-Hampton, M. and Vavrina, C.S. (2002). Worm castings: An alternative to sphagnum peat moss inorganic (_Lycopersicon esculentum Mill_) transplant production. pp 105-113. In: F.C Michel, R.F..Rynk and H.A.J. Hoitink (Eds). roc. Intl.Symp.Composting and Compost Utilization. Columbus, OH, June 6-8.

Parle, J.N. (1963). Micro-organisms in the intestines of earthworms. _J. Gen. Microbiol.,_ 31:1–11.

Parmelee, R. W. and Crossley D.A. Jr. (1988). Earthworm production and role in the nitrogen cycle of a no-tillage agroecosystem on the Georgia piedmont. *Pedobiologia,* 32(5): p.355-361.

Pashanasi, B, Lavelle, P., Alegre, J. and Chapenetier, F. (1996). Effect of the endogeic earthworm, *pontiscolux corethrurus* on soil chemical characteristics and plant growth in a low-input tropical agro ecosystem. *Soil Biology & Biochem.,* 28 (6), 801-808.

Piearce T.G. 1978. Gut contents of some lumbricid earthworms. *Pedobiologia,* 18: 153-157.

Ranganathan and Cristopher (1994). Effect of vermicompost on soil fertility and response of horticultural crops. *Crop Research,* 8 (3): 453-456.

Ranganathan, LS, Vinotha, S.P. (1998). Influence of pressmud on the enzymatic variations in the different reproductive stages of *Eudrilus eugineae. Current Science,* 74:634-635.

Rani, R. and Srivastava, O.P. (1997). Vermicompost: a potential supplement to nitrogenous fertilizer in rice nutrition. *Int. Rice Res. Notes.,* 22 (3):30-31.

Rao, K. J. and Shantaram, M. V. (1994). Heavy metal pollution of agricultural soils due to application of garbage. *Ind. J.Environ.Health.,* 36 (1): 31-39.

Ross D.J., Cairns A. 1982. Effects of earthworms and ray grass on respiratory and enzyme activities of soil. Soil Biol. Biochem. 14:583-586.

Ruz-Jerez, B.E., Ball, P.R. and Tillmann, R.W. (1992). Laboratory assessment of nutrient release from a pasture soil receiving grass or clover residues, in the presence or absence of *Lumbricus rubellus* or *Eisenia fetida. Soil Biol. Biochem.,* 24: 1529-1534.

Saciragic, B. and Dzelilovic, M. (1986). Effect of worm compost on soil fertility and yield of vegetable crops cabbage leeks and sorghum hybrid yield. *Agrohemija,* 3: 343-351.

Sagar, D.V., Naik, S.N. and Vasudevan, P. (2002). Vermicomposting of kitchen waste and its effect on the quality and quantity of essential oil of *Ocimum sanctum* L. Proc Intl Composting and Compost Sci Symposium, Columbus, Ohio: CD Rom.

Santra, S. K. and Bowmik, K.L. (2001). Vermiculture and development of agriculture. *Yojna.* 43-45.

Senesi, N., Saiz-Jimenez, C. and Miano, T.M. (1992). Spectroscopic characterization of etal-humic acid-like complexes of earthworm-composted organic wastes. *The Science of the Total Environment,* 117/118: 111–120.

Sharma, Kaviraj (2003). Municipal solid waste management through vermi-composting employing exotic and local species of earthworms. *Biores. Technol.,* 90: 169–173.

Shivsubramanian, K. and Ganeshkumar, M. (2004). Influence of vermiwash on the biological productivity of marigold. *Madras Agric. J.,* 91 (4-6) : 221-225.

Shuster, W.D., Subler, S. and McCoy E.L. (2000). Foraging by deep-burrowing earthworms degrades surface soil structure of a fluventic Hapludoll in Ohio. *Soil Tillage Res.,* 54: 179-189.

Singh, Janardan and Rai, S.N. (1997). Earthworm farming and vermicomposting: A boon for sustainable Agriculture. *J. Soil Biol. Ecol.,* 17(1):65-72.

Stockdill, S.M.J. and Cossens, G.G. (1966) .The role of earthworms in pasture production and moisture conservation. *Proceeding of the New Zealand Grassland Association,* 168-183.

Subler, S., Edwards, C.A. and Metzer, J. (1998). Comparing vermicomposts and composts. *Biocycle,* 39: 63-66.

Thimmaiah A. 2001. Studies on Biodynamic system and vermitechnology for sustainable agriculture. Ph.D Thesis, IIT Delhi.

Tiwari, S.C., Tiwari, B.K. and Mishra, R.R. (1989). Microbial populations, enzyme activities and nitrogen–phosphorous–potassium enrichment in earthworm casts and in the surrounding soil of a pineapple plantation. *Biol. Fertil. Soils,* 8: 178–182.

Todkari, A.A. and Talashilkar, S.C.(2001). Effect of vermiwash prepared by two methods on growth characteristics, yield and nutrition of three following plants. Souvenir and Abstracts. A paper presented in Silver Jubilee Celebrations of the Indian Society of Soil Biology and ecology and VII National symposium on soil Biology and Ecology held at U.A.S, Bangalore during Nov.7-9, 2001:pp 97.

Tomati, U., Grappelli, A. and Galli, E. (1988). The hormone-like effect of earthworm casts on plant growth. *Biology and Fertility of Soils,* 5: 288–294.

Van Rhee, J.A. (1969). Inoculation of earthworms in a newly drained polder. *Pedobiologia,* 9: 128-132.

Van Rhee, J.A. (1971). Some aspects of the productivity of orchards in relation to earthworms activity. *Annal of Zoology and Ecology* (Special publication) 4: 99-108.

Vasudevan, P. (2001): Status reports on solid waste management by vermicomposting, UNICEF sponsored project.

Werner, M. and Cuevas, R. (1996). Vermiculture in Cuba. Biocycle, vol. 37, JG Press, Emmaus, PA., pp. 61–62.

Heavy Metal Contamination of Soil and Groundwater due to E-waste Handling in Mandoli Industrial Area of Delhi, India

**Sirajuddin Ahmed[1], Rashmi Makkar Panwar[2]
and Anubhav Sharma[2]**

[1]*Department of Civil Engineering, Jamia Millia Islamia, New Delhi-110025*
[2]*G.B.Pant Institute of Technology, DTTE, New Delhi, India*

Abstract

This paper reveals the magnitude of heavy metal contamination of soil and ground water in and around an unauthorized E-waste recycling site in Delhi. Though unsafe and unorganized e-waste handling is now legally banned in Delhi, still the informal sector is actively involved in carrying out dismantling, extraction and disposal of E-waste at certain places at considerably large scale. The leachate produced by these recycling units contains a large amount of heavy metals which are likely to pollute the groundwater and soil adjoining the recycling sites. The E-waste contamination at such sites is evaluated in this study by monitoring the potential contaminants at a number of specific monitoring points. The soil and underground water quality is checked for the presence of heavy metals around e-waste recycling and dumping site using Atomic Absorption Spectrometry (AAS). The standard values as per central ground water board are taken as reference values for water, and agricultural soil in Britain as for soil at the e-waste site. It is clearly evident from the results that the groundwater and soil in and around these sites have been found contaminated by heavy metals like lead and copper to a great extent.

Keywords: Heavy metals, contamination, groundwater, soil.

Introduction

The Waste Electrical and Electronic Equipments (WEEEs) contain several substances, many of which are toxic in nature and could be hazardous for environment, specially soil and water, whereas precious metals including gold, silver, palladium, tantalum, platinum etc are only present in traces in WEEEs. While recycling in automatic methods, precious metals are lost in bulk of other less valuable materials (Chatterjee and Kumar, 2009). Dissolved metals are considered to be the most mobile and thus reactive. These are bio available fractions in an aquatic system and are cause of major concern(Wong *et al.*, 2007).

In this study various sites of e-waste handling and recovery were visited and it was found that metal recovery is still taking place in Mandoli, New Delhi, India. Unauthorized e-waste handling is banned but it was noticed that at some places waste handling operations remain unchecked and covertly carried out banned activities. The large quantity of e-waste which is dumped after recovery in these areas portrays the real scenario. The samples of soil and water were collected and tested for heavy metal contamination from the identified sites .The testing was done for copper, lead, cadmium, nickel, chromium, and zinc.

Effects of heavy metal contamination

The metals that are considered as heavy are those with a density greater than a certain value, usually 5 or 6 g/cc (Wild, 1996). The e-waste contamination can adversely affect fertility of soil whereas it can render water unfit for consumption as heavy metals can leach into soil and water when e-waste is irresponsibly dumped. Increased heavy metal content negatively affects soil microbial population, which may have direct negative effect on soil fertility (Ahmad *et al.*, 2005). Heavy metals released from salvaging useful materials and from open burning could pollute air, soil and water. Plants can take up these metals from roots, transport them upwards to their shoots, and finally accumulate them inside their tissues (Luo *et al.*, 2006). Heavy metals not recovered during WEEE treatment and residual auxiliary substances like mercury and cyanide can leach through the soil after disposal of effluents and form inorganic and organic complexes within soils (Sepulveda *et al.*, 2010). Leaching of heavy metals through the soil prevents plants from absorbing their essential nutrients (Mancuso and Green, 2010), such as potassium, calcium, magnesium, and nitrates. Also, it can disturb the ecological balance of the region by disrupting the growth of microorganisms.

Heavy metals cadmium, zinc, lead and chromium can lead to human poisoning when consumed in drinking water. Consumption of heavy metals cause irregularity in blood composition, badly effect vital organs such as kidneys and liver (Khan *et al.*, 2011) apart from damaged or reduced mental, central nervous function and lower energy level. The long term consumption of these metals result in physical ,muscular, neurological degenerative processes that cause Alzheimer's disease, Parkinson's disease, muscular dystrophy and multiple sclerosis (Mohod and Dhote, 2013).

Previous studies

Similar study has been carried out in the largest E-waste dumping site in Alaba International market in Lagos, Nigeria, in which concentrations of heavy metals mentioned above were identified in the soil at the site (Olafisoye *et al.*, 2013). In that study, concentration levels of Cadmium, Chromium, Lead, Nickel and Zinc in water, soil and plants were measured by Atomic Absorption Spectrometry (AAS) using wet digestion method. It was found that concentration levels of heavy metals in consideration exceeded the permissible levels in soil at the site. In China, soils at sites where e-waste is burned (Luo *et al.*, 2010) was analyzed for quantifying heavy metal contamination levels and soil samples of former incineration sites were found to have exceedingly high levels of Cd, Cu, Pb and Zn. E-waste recycling operations cause appreciable hazards to adjoining soils. In Oman, testing was done for surface and ground water contamination (Al Raisi *et al.*, 2014) at unlined leachate sites of Al Amerat and samples were analyzed for concentration levels of heavy metals in drinking water. Metals like Nickel, Cobalt and lead were found to be in exceeding amount in waste and drinking water. In India, Toxics Link has conducted study at E-waste recycling site in Loni and Mandoli areas in Delhi (Toxics Link, 2014) and has tested for heavy metal contamination of soil and water. Recycling practices at these sites were studied and their effect on soil and water concentrations was analyzed. Soil samples from these sites suggest there has been significant concentration of heavy metals and soil characteristics have changed and hence the correlation between recycling practices at these sites and soil characteristics were established.

Need of study

In India, only a few landfill sites are available for environmentally sound disposal of E-waste. And severe environmental hazards are associated with careless and irresponsible dumping of E-waste that contains several hazardous substances. The entrepreneurs engaged in recycling lack proper technical knowledge and do not have adequate means to handle the increasing quantities nor the expertise for certain recovery processes (Ha *et al.*, 2009). These substances pose serious adverse impacts on soil and water adjoining the site. Not even households, but several industries are involved in such illegal practices and contribute to environmental hazards. The illegal dumping sites are a major issue of concern for India. Quantification of hazardous material contaminating the soil and water at the landfill and recycling sites is significant for environmental organizations and legal authorities in India to pressurize the recyclers for using alternative measures of recycling and disposal. Also, it helps keeping a check on the illegal dumping of E-waste and make the producers more responsible towards safe disposal of their end of life products. Moreover, the in-complete data on generation and disposal of hazardous substances increase the worry. It is presumed that about 10 to 15 percent of wastes produced by industry are hazardous and the generation of hazardous wastes is increasing at the rate of 2 to 5 percent per year (Trehan, 1992). But, reliable data and exact assessment on quantity of various hazardous wastes generated is not available as yet. Therefore, scientific disposal of E- waste has become a major challenge in India.

WEEE is one of the most complex waste streams requiring proper management. To meet the challenge associated with disposal of waste from electrical and electronic products, the Central Pollution Control Board in India has issued guidelines for environmentally sound management and handling of e-waste that came in effect from 1st May 2012 (MoEF, 2010).

Material and Methods

Study area

The selected site for this project was Mandoli industrial area in East Delhi as it is one of the hubs of unorganized and uncontrolled e-waste handling (Gidarakos *et al.*, 2012) and dumping in Delhi. The area of interest can be located as having western borders approaching the river Yamuna and North and Eastern sides are densely populated zones. The area is swamped with small one or two room unauthorized E-waste recycling units (MMA, 2008) in which majority of the population finds employment. The informal e-waste recycling sites discharge their effluents into open lands in the absence of drains and solid waste is disposed by open burning. In Mandoli, there are 6 companies running since year 2000 which recover copper by burning Printed Wire Boards and nearly 3000 kg of PWBs are burnt per day (Malik, 2004). Also, large dumps of waste of electronic products can be seen lying openly on the roadsides, as witnessed during personal visits. Ground adjoining to farm land is also used for disposal of e-waste. Thus the effluents are directly discharged into land directly impacting soil and ground water. The area is densely populated and the use of underground water is very common by the residents of this area and surrounding localities. The scope of this study is restricted to soil and water contamination by heavy metals released as e-waste disposal after recycling and extraction.

Soil and water sampling

All the soil and water samples were collected in December 2013.

Soil samples: 10 top soil samples from ground level and 10 sub-surface soil samples were taken from about 60 cm deep soil. The collection of samples is done from various locations in Mandoli. Reference soil samples were taken from the locality 5 Km. away from e-waste handling units. All the samples were collected using stainless steel spade and were put in plastic containers. Samples of underground water were also collected to assess the effect of heavy metals released from e-waste recycling unit on the groundwater. Water samples were collected from hand pumps located in the vicinity of dumped e-waste and reference samples were collected from the areas 5km away from active sites for comparison of the heavy metal content of all the samples. Water samples were taken in clean plastic one liter bottles and were tested using Atomic Absorption Spectroscopy after digesting with HNO3. A conventional wavelength-dispersive X-ray fluorescence spectrometer was used for AAS (Kawai *et al.*, 1998), as it has high energy resolution and short measuring time.

Sample analysis

Soil samples were collected and investigated for presence of heavy metals and standard procedures as indicated in Table 1 were used for soil analysis.

Table 1: Standard procedure used for sample analysis

S.No	Parameter	Method of Analysis	Unit	Procedure	Analyzed Substance
1	Copper	APHA 3120-B	mg/kg	Digestion with HNO_3, AAS (Atomic Absorption Spectrometry	Soil and water
2	Lead	APHA 3120-B	mg/kg	Digestion with HNO_3, AAS (Atomic Absorption Spectrometry	Soil and water
3	Cadmium	APHA 3120-B	mg/kg	Digestion with HNO_3, AAS (Atomic Absorption Spectrometry	Soil and water
4	Nickel	APHA 3120-B	mg/kg	Digestion with HNO_3 ,AAS (Atomic Absorption Spectrometry	Soil and water
5	Chromium	APHA 3120-B	mg/kg	Digestion with HNO_3 ,AAS (Atomic Absorption Spectrometry	Soil and water
6	Zinc	APHA 3120-B	mg/kg	Digestion with HNO_3 ,AAS (Atomic Absorption Spectrometry	Soil and water

Results and Discussion

The soil samples collected from the E-waste site and reference site after investigation yielded the results as shown in Table 2. The concentration levels of heavy metals in Top soil and Sub soil samples of the E-waste and reference sites are indicated below. The values indicated are the average values for metal concentration. The standard values for metal concentrations in soils are those for agricultural soils in Great Britain (Davies and Ballinger, 1990).

Table 2: Heavy metal concentrations (mg/kg) in soil samples of different sampling sites

Site	E-waste site		Reference site		Agricultural soil (in Great Britain)
Heavy Metal	Top soil	Sub soil	Top soil	Sub soil	Standard
Cu	238.23	73.04	8.39	0.58	100
Pb	298.10	183.54	12.50	0.43	100

Site	E-waste site		Reference site		Agricultural soil (in Great Britain)
Cd	47.77	19.16	0.26	0.00	3
Ni	41.44	40.14	7.66	0.00	50
Cr	145.18	80.53	6.99	0.1	50
Zn	174.83	65.11	9.69	0.30	300

Soil Contamination

The mean heavy metal concentrations of soil samples collected from e-waste handling sites and reference sites are shown in Table 2. To evaluate the extent of heavy metal contamination in soil, the concentrations were compared with reference soil samples and soil guidelines given by standards for agricultural soils in Great Britain. The average concentrations were copper 238.23mg/kg, lead 298.10mg/kg, cadmium 47.77mg/kg, nickel 41.44mg/kg, chromium 145.18mg/kg and Zinc 174.83mg/kg in top soil samples (see figure 1). The average concentration of Copper, Lead, Cadmium and Chromium topsoil samples from E-waste site are far above the range for standards of agriculture soil and exceed the permissible limits. The concentrations of all these heavy metals are significantly higher than the reference site concentrations also. The average copper concentration is nearly 30 times as compared to reference sites top soil samples and nearly 120 times for sub soil samples. Copper is associated with organic matter, oxides of iron and manganese, silicate clays and few other minerals. It builds up in the surface of contaminated soils showing virtually no downwards migration (Parth *et al.*, 2011). The major reason of the high copper concentrations in all the soil samples is caused due to copper extraction from printed wiring boards, which is one of the major activity carried out in the area.

Printed circuit boards, wires and lead batteries are treated at sites of selection, hence it becomes important to study the impact of lead on soil and water in surrounding area Lead is used in electric solder, primarily on printed circuit board. The average lead concentrations were found to be 298.10 mg/kg in top soil samples and 183.54mg/kg in sub soil samples of e-waste handling sites which are almost three and two times respectively higher than the standard concentrations. Reference soil concentrations are considerably very low as compared to sites concentrations. Lead (Pb) in soil exists in the +2 oxidation state .As soil pH raises; Pb^{2+} ion becomes less soluble under oxidizing conditions. Metallic lead has been used in electric solder, commonly as an alloy with tin and lead compounds have been commonly used as stabilizers in PVC (Polyvinyl chloride) formulations (OECD, 2003).

Average concentrations of Cadmium in top soil samples is 16 times higher than that of agricultural standards and 6 times higher than in subsoil samples ,whereas the reference site samples have negligible cadmium concentrations at an average of 0.26 mg/kg. Cadmium and its compounds are used in a number of applications within electrical and electronic products (Shankar *et al.*, 2005).

The average concentrations of nickel in top soil and sub soil are nearly same at 41.44 mg/kg and 40.14 mg/kg .The average values are within the threshold limit of 50mg/kg, however the average concentrations level exceeds the average reference values.

Average concentrations of chromium in topsoil samples is 145.18 mg/kg which is approximately 3 times higher than the standard limit and 20 times higher than the reference value. Average concentration in sub soil samples is also 1.5 times than standard limits and 800 times than the reference site subsoil average concentration. Waste from lead –chromium batteries, coloured polythene bags, discarded plastic materials and empty paint containers are said to be huge source of chromium (Matthews, 1996).

Average concentration of zinc was found to be within the limits of standards for agricultural soil.

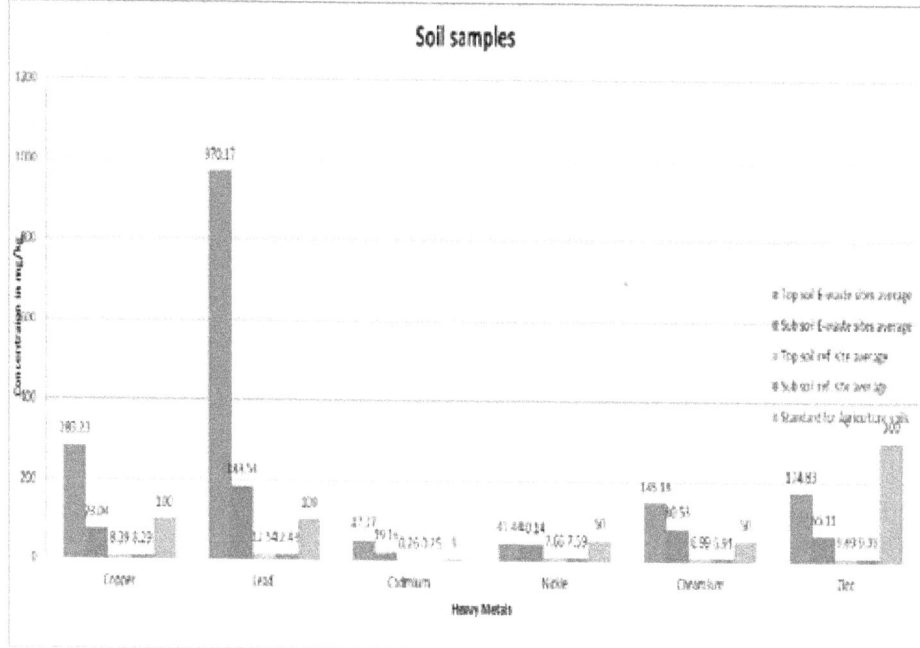

Fig. 1: Heavy metal concentrations in soil samples

Heavy metal concentrations (mg/kg) in soil samples of different sampling sites

To evaluate the extent of heavy metal contamination in soil, the concentrations in collection sites samples were compared with reference soil samples and soil guidelines given by standards for agricultural soils in Great Britain. Table 3 shows the average values of heavy metals composition evaluated for Top soil samples of E-waste sites and reference sites. The comparative analysis is depicted in table below as follows:-

Table 3: Heavy metal concentrations (mg/kg) in Top soil samples of different sampling sites

Heavy Metal	E-waste site	Reference site	Standard (Agricultural soil in Great Britain)
Cu	238.23	8.39	100
Pb	298.10	12.50	100
Cd	47.77	0.26	3
Ni	41.44	7.66	50
Cr	145.18	6.99	50
Zn	174.83	9.69	300

The average concentrations were copper 238.23mg/kg, lead 298.10mg/kg, cadmium 47.77mg/kg, nickel 41.44mg/kg, chromium 145.18mg/kg and Zinc 174.83mg/kg in top soil samples. The average concentration of Copper, Lead, Cadmium and Chromium found in topsoil samples from E-waste site are far above the range for standards of agriculture soil and exceeds the permissible limits. The concentrations of all these heavy metals are significantly higher than the reference site concentrations also. The average copper concentration is nearly 30 times as compared to reference sites top soil samples and the average lead concentrations were found to be 298.10 mg/kg in top soil samples, which is almost three times higher than the standard values.

Water contamination

The results of water samples collected are as shown below in table 4.

Table 4: Heavy metal concentrations (mg/kg) in water samples of different sampling sites

Heavy metals	E-waste site	Reference site	Standard*
Cu	1.465	0.18	0.05
Pb	1.25	0.0075	0.05
Cd	0.28	0.00	0.01
Ni	0.29	0.003	0.05
Cr	0.008	0.008	0.05
Zn	0.016	0.016	5.00

Drinking water standards as per central ground water board (Shekhar., 2012)

The average concentration of copper in water sample is 29 times higher than the water standards and 8 times higher than the reference water levels. The average lead concentration is 25 times the threshold value. Cadmium concentration in reference samples is 0 whereas the average value of cadmium is 0.28 mg/.lit. .The threshold limit for drinking water is 28 times lower than this value making this water highly unsafe for drinking. Average Nickel concentration is 0.29mg/lit which is exceeding the threshold value of 0.05 mg /lit. The average concentration of chromium is 0.83mg/lit and the reference water samples average is far less at 0.008 mg/lit (see figure 2). This value also exceeds the drinking water limits. It has been observed that the concentrations of all the heavy metals found in water samples were found to be more than the permissible limits and the reference water sample concentrations.

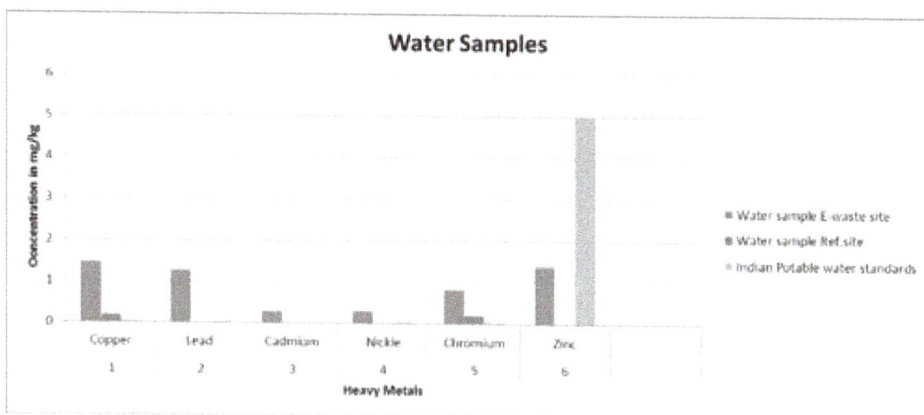

Fig. 2: Heavy metal concentrations (mg/kg) in water samples of different sampling sites

3.3 Statistical analysis

There were significant associations between cadmium and copper ($P<0.05$) and chromium and lead ($P<0.05$) in soil samples. The correlation coefficient is also significant for cadmium and copper ($P<0.05$) and chromium and nickel ($P<0.05$) in water samples. The findings for correlation coefficients of metals in soils and water are shown in table 10 and table 11 respectively. This suggests that elevated cadmium concentrations are associated with higher values of copper concentrations in soil as well as water samples. Significant correlations between metals indicate a common source. The results of testing for various samples are compiled in tables 5 and 6.

Table 5: The correlation co-efficient of metals in soils

	Cu	Pb	Cd	Ni	Cr	Zn
Cu	1					
Pb	-0.035	1				

	Cu	Pb	Cd	Ni	Cr	Zn
Cd	0.895*	-0.203	1			
Ni	0.569	0.0435	0.604	1		
Cr	0.343	0.766*	0.276	0.063	1	
Zn	0.057	-0.014	-0.058	-0.065	0.125	1

Correlation is significant at P<0.05

Table 6: Correlation co-efficient of metals in water

	Cu	Pb	Cd	Ni	Cr	Zn
Cu	1					
Pb	-0.096	1				
Cd	0.977*	-0.266	1			
Ni	0.682	0.155	0.578	1		
Cr	0.611	-0.048	0.568	0.885*	1	
Zn	-0.109	0.468	-0.222	0.432	0.048	1

Correlation is significant at P<0.05

Conclusion

This study shows that the average concentrations of copper, lead cadmium and chromium of all topsoil samples fall out of the range for standards of agriculture soil and exceeds the permissible limits and are significantly higher than the reference site concentrations. Nickel concentration is within limits in soil samples but in water it is about 5 times higher than potable water standards. Copper concentration is more than 2 times higher than the standard limit in top soil samples and alarming at a level of 1.46 mg/kg in water samples as Indian potable water standards are 0.05mg/kg. Average Lead concentration in soil is approximately 3 times higher than the standard limit and in water samples it is 25 times higher than the potable water limits. Average cadmium concentration in soil is 16 times higher than standard limits and in water it is 29 times higher than that of Indian potable water standards. Chromium concentration is about 3 times higher than the standard limit and 16 times higher than the potable water limits. Zinc concentration is within prescribed limits for soil and water samples.

Testing and analysis of water samples clearly indicate the presence of heavy metals in all the underground water samples which also exceeds the limits for drinking water. The water of these areas is not suitable for drinking. The findings clearly indicate the contamination of soils due to heavy metals released during processing of e-waste and it provides useful baseline information on e-waste contamination for soil. The emission of heavy metals and the risk for the health by direct exposure or by bioaccumulation and geo-accumulation is of great concern. In addition to consumption of water local inhabitants are also susceptible to acid fumes, ingestion of heavy metal contaminated dust and toxic gases on account of e-waste recycling practices carried out around this area. These findings point towards the requirement of better monitoring and enforcement of laws to control informal and improper handling.

References

Ahmad, I., Hayat, S., Ahmad, A., Inam, A. and Samiullah, I. (2005). Effect of heavy metal on survival of certain groups of indigenous soil microbial population. *Journal of Applied Science and Environment Management*, 9: 115–121.

Al Raisi, SAH, Sulaiman, H. and Suliman, F. (2014). Assessment of Heavy Metals in Leachate of an Unlined Landfill in the Sultanate of Oman. *International Journal of Environmental Science and Development*, 5(1): 60-63.

Chatterjee, S. and Kumar, K. (2009). Effective electronic waste management and recycling process involving formal and non-formal sectors. *International Journal of Physical Sciences*, 4(13): 893-905.

Davies, B.E. and Ballinger, R.C. (1990). Heavy metals in soils in north Somerset, England, with special reference to contamination from base metal mining in the Mendips. *Environmental Geochemistry and Health*, 12(4): 291-300.

Gidarakos, E., Basu, S. and Rajeshwari, K.V. (2012). E-waste recycling environmental contamination: Mandoli, India. 2012. *Proceedings of the ICE - Waste and Resource Management*, 165(1): 45-52.

Ha, N.N., Agusa, T. and Ramu, K.(2009). Contamination by trace elements at E-waste recycling sites in Bangalore, India. *Chemosphere* 76: 9-15.

Kawai, J., Hayashi, K., Okuda, K. and Nisawa, A. (1998). X-ray absorption spectroscopy using X–ray fluorescence spectrometer. *The Rigaku Journal*, 15(2): 33-38.

Khan, S. A., Zahoor, U.D. and Ihsanullah, Zubair A (2011). Levels of selected heavy metals in drinking water of Peshawar city. *International Journal of Science and Nature*, 2(3): 648-52.

Luo, C., Liu, C. and Wang, Y.(2010). Heavy metal contamination in soils and vegetables near an e-waste processing site, south China. *Journal of Hazardous Materials*, 186(2011): 481-490.

Luo, C, Shen, Z., Lou, L. and Li, X. (2006). EDDS and EDTA-enhanced phytoextraction of metals from artificially contaminated soil and residual effects of chelate compounds. *Environmental Pollution*, 144 (3), 862-71.

Malik, R. (2004). Risk Assessment of E-waste burning in Delhi, India. *Diploma Thesis of Stefanie Steiner*, Zurich.

Mancuso, A. and Green, T. (2010). The Effect of the Accumulation of Heavy Metals in Soil on the Growth of Vegetation in the Long Island Solar Farm. Science Undergraduate Laboratory, Office of Science, Department of Energy, Siena College.

Mathews, G. (1996). PVC: production properties and uses. The Institute of Materials, London.

MMA (Ministry of Minority Affairs) (2008). Baseline Survey of North-East District, NCT Delhi. Jamia Millia Islamia, New Delhi, India.

MoEF (Ministry of Environment and Forest) (2010). Modified draft notification E-waste (Management and Handling) rules PartII section 3 sub-section ii. Gazette of Govt of India.

Mohod, C.V. and Dhote, J. (2013). Review of heavy metals in drinking water and their effect on human health. *International Journal of Innovative Research in Science, Engineering and Technology*, 2(7): 2992-96.

OECD(2003). Technical guidelines for environmentally sound management of specific waste streams used and scrap personal computers .Working Group on Waste Prevention and Recycling. www.oecd.org/env/waste/39559085. pdf.

Olafisoye, O.B., Adefioye, T. and Osibote, O.A. (2013). Heavy Metals Contaminations of water, soil and plants around an electronic waste dumpsite. *Polish Journal of Environmental Studies*, 22(5): 1431-39.

Parth, V., Murthy, N.N. and Saxena, P.R. (2011). Assessment of heavy metal contamination in soil around hazardous waste disposal sites in Hyderabad city (India): natural and anthropogenic implications. *Journal of Environment Research and Management* 2(2):27-34.

Sepúlveda, A., Schluep, M., Fabrice, G., and Renaud, F.G. (2010). A review of the environmental fate and effects of hazardous substances released from electrical and electronic equipments during recycling: Examples from China and India. *Environmental Impact Assessment Review*, 30(1): 28-41.

Shanker, A.K., Cervantes, C., Tavera, H.L. and Avudainayagamd, S. (2005). Chromium toxicity in plants. *Environment International*, 31:739-753.

Shekhar, S., Purohit, R.R. and Kaushik, Y.B. (2012). Groundwater Management in NCT Delhi. Central Ground Water Board, Govt of India, New Delhi.

Kabata-Pendias, A. and Pendias, H. (1992), Trace Elements in Soils and Plants, 2nd ed. CRC Press, Boca Raton, FL.

Toxics Link (2014). A report on Impact of E-Waste Recycling on Water and Soil, New Delhi, India.

Trehan, N.C. (1992). Environmental aspects of hazardous wastes disposal in India. *Environmental impact assessment of developing countries*,124-130.

Wild, A. (1996). Soils and Environment: An Introduction. Cambridge University Press, Cambridge: 290.

Wong, C.S., Duzgoren-Aydin, N.S., Aydin, A. and Wong, M.H. (2007). Evidence of excessive release of metals from primitive e-waste processing in Guiyu,China. *Environmental Pollution*, 148(1): 62-72.

Riparian Vegetation Functions and Ecological Services: Interlinking Soil Environment Perspectives

M.K. Jhariya* and D.K. Yadav

Department of Farm Forestry, Sarguja University, Ambikapur-497001 (C.G.), INDIA

Abstract

Riparian area are one among the most diverse, complex and dynamics ecosystems on earth. These vegetation have substantial and diverse economic, environmental and ecological consequences. These ecosystems provide food and habitat for unique plant and animal species, and are essential to the mitigation and control of non-point source of pollution. The relationship of riparian vegetation with environment and ecosystem are poorly understood for various rivers systems of India in general and its function and ecological services particular. Many plant and animal species depend on the distinctive habitat of riparian vegetation area, which include elements of both terrestrial and aquatic ecosystems. These areas improve habitat quality by facilitating various ecological and environmental functions and influence the surrounding landscape. Characteristics riparian vegetation affect bank stability, erosion, channel morphology, stream flow, water temperature, inputs of sediment, nutrients and organic litter, which sustain water quality and aquatic food webs. Therefore, to achieve these benefits in a sustainable manner, efforts to restore riparian zones must occur at appropriately large spatial scales and other drivers of degradation must be addressed.

Keywords: Riparian vegetation, soil environment, environmental consequences, restoration

Introduction

The total environmental conditions, resource availability, cultural diversity and identity of the human being are very much related to the physical geography of landscape. This fact becomes very decisive in subtropical subcontinent. Human impact on our planet is immense, and extends to all biomes and ecosystems which seems to be impressive and alarming. According to FAO (2010) about 13 mha of forest are converted into different form or land use each year, which is approximately 0.13% of total forest cover of the world. UNEP (2014) revealed that at present the emissions of GHGs have accelerated from 200 MT to 289 billion tonnes/year since 1850. This alarming phenomenon of climate change and global warming has tend to influence and reduce polar sea ice, radically alter the arctic ecosystems (Smol *et al.*, 2005) and continues to fluctuate weather patterns and different environmental segments throughout the world.

Water use, catchment modification and river pollution continue to increase and freshwater ecosystems are experiencing great challenges even greater biodiversity loss than terrestrial ecosystems (Dudgeon *et al.*, 2006). Only 0.16% of the land area is unaffected globally by any anthropogenic interferences (Vorosmarty *et al.*, 2010). Global warming, climate change, water quality and quantity, food insecurity and biodiversity loss are among the serious environmental issues throughout the world among the scientific community (Diamond, 2005). Pressure placed on the world's natural resources by growing population has led to severe issues, problems and challenges related to environment and ecosystem (Foley *et al.*, 2005). With the rise of global population it tends towards change of land use for increase the production and productivity from agricultural landscape which leads to conversion of fragile natural landscape under agricultural land-use system (Foley *et al.*, 2005).

Now-a-days the importance of the riparian vegetation is greaty acknowledge and understood by various scientific communities of the world. This happened not because of scientific reasons related to the development of riverine landscape ecology (Tockner *et al.*, 2002), but due to riparian areas being one of the key concerns towards sound management of river systems. Now, these areas are seriously degraded by different biotic factors and their activities. Furthermore, the scientific understanding of their hydrological, environmental and ecological aspects along with functioning is greatly needed for their restoration implication, preparing restoration action plan and strategies for their scientific management and conservation especially in Indian river systems.

Riparian: Meaning and Concept

In the early of 1800's the United States was initially used the word "Riparian" as a legal term (Ortega Klett, 2002). It is unclear when scientists first adopted the term "riparian" to describe the areas adjacent to streams, rivers, lakes and near to water bodies. This term started appearing in the scientific literature in the 1970's (Baker, 2002). In the journey for last three to four decades, our understanding of the importance of environmental, ecological and hydrological processes in riparian areas has increased (Baker, 2002).

The word 'Riparian' derived from the Latin word '*Ripa*', which means river bank, streams, wetlands, pond or lake of the surrounding landscape (Tabacchi *et al.*, 1990; Goebel *et al.*, 2003). These ecosystems have direct impact on aquatic, terrestrial ecosystem and wildlife habitat, and it also known as stream side forests or gallery forests (Brinson, 1990). These can act as a mosaic of micro-habitats with the coexistence of rich floral diversity and species richness (Swanson *et al.*, 1982). Riparian landscapes are rare habitat and represent small fraction of the total geographical area of the world's landscapes but are one among the highly threatened ecosystems (Hynes, 1970).

The grasses, shrubs and trees growing in these sensitive areas are called as riparian vegetation. Functionally, riparian zones can be defined as "3-dimensional zones of interaction between terrestrial and aquatic ecosystems" (Sedell *et al.*, 1991). Riparian vegetation generally consists of three layered organization of canopy trees, middle stratum of shrubs and woody climbers and herbaceous as ground strata. The zonation of above ground vegetation of the riparian ecosystems rely on the site conditions and regional climate. Tree stratum considered as the most significant component, due to the keystone nature of riparian forests (Minore and Weatherly, 1994; Pettit and Froend, 2001).

Bilby (1988) repoted that the size of riparian zone is proportionate to size of stream and site topography; steep slopes characteristic of small streams may limit the vegetation development in riparian areas, whereas aquatic systems with less extreme topography exhibit larger riparian boundaries. Conversely, the consequence of the riparian zone on the aquatic system decreases as size of stream increases (Agee, 1988; Bilby, 1988). Anthropogenic pressures tend to increase surface runoff in riparian systems, remove protective cover and modify the water flow through aquatic ecosystems (Manci, 1989).

Riparian Vegetation Structure and Ecology

The composition, structure, ecological features and function of riparian vegetation have very unique attributes and dimension. These are dependent and regulated by biotic and abiotic component of the environment. Malanson (1993) discussed three paradigms that may be used to explain the ecology and high structural variability of riparian sites. These paradigms are the individualistic hypothesis of plant associations, the intermediate disturbance hypothesis and the competition hypothesis of niche relations.

⬦ According to the individualistic hypothesis, water gradient may be used to depict the species distribution and communities within a riparian site, and it is often more variable and does not always follow that pattern and that the relationship is not always linear.

⬦ The second paradigm used to explain vegetation distribution in riparian areas is the intermediate disturbance hypothesis (Connell, 1978). This hypothesis states that species diversity may be greatest when disturbances are intermediate in area, intensity and frequency. In many riparian areas, vegetation is thought to be controlled by flooding (Malanson, 1993). Flooding frequency and severity will control plant communities by dictating species colonization, establishment and survival. Grazing can also be viewed as an ecological disturbance.

⬦ The third paradigm considered by Malanson (1993) is the competition hypothesis of niche relations. Competition within and between species may control the distribution of plant communities along an environmental gradient. According to this theory, a species is excluded from those portions of the gradient where it is least adapted to compete and expand when species competition ceases (Malanson, 1993).

Although strict application of each of the above-mentioned concepts has obvious flaws and complications, the underlying principles of each theory deserve consideration when studying and managing riparian ecosystems. However, a combination of all three paradigms probably best explains the distribution and heterogeneity of riparian vegetation. The kind, amount and vegetation distribution on riparian area are largely controlled by soil water (Leonard *et al.*, 1992).

Status and Challenges

Riparian ecosystem represent the small fraction of global concern but have a rich floral and faunal diversity and one among the highly vulnerable ecosystem. Human civilization have its overall impact and accelerating transformation and alteration of the major rivers systems of Asia such as Indus, Ganges and Yangtze (Dudgeon, 2000) and are now categorized as threatened ecosystems (Johnsingh and Joshua, 1989; Dudgeon, 1992) due to the biodiversity loss and species richness. In India, the changes on phytodiversity of riparian forests are alarming and it found under threats due to various disturbances regimes such as deforestation, overgrazing and land reclamation etc. (Gopal, 1988). Riparian forests adjoining stream and river banks have been almost entirely eliminated outside the protected areas (Gadgil, 2004). The major challenges related to riparian forests includes:

◈ Severely compromised and has been subjected to severe disturbances. In many region the riparian zone is extremely narrow or in some cases completely cleared or damaged.

◈ Presence of exotic and invasive alien species along the catchments. Many of these non-native species are found extensively in these belts and are not limited to the riparian zones, although they appear to proliferate in these habitats and adjoining forest areas.

◈ Changes and loss of biodiversity and alteration in health of the water bodies.

◈ Global warming and climate change altering the water flow, aquatic ecosystem and riparian vegetation.

Riparian Vegetation: Functions and Services

Riparian vegetations have specific functions and services in the landscape as they play vital role in both aquatic and terrestrial ecosystem (Malanson, 1993). They are distinctly differ from the surrounding lands because of unique soil and vegetation characteristics. The interfaces in riparian zones possess physical and chemical attributes, biotic processes and material flow processes (Vannote *et al.*, 1980).

Riparian areas can serve many functions such as filtering pollutants, stabilizing stream banks, storing surface water and sediment, maintaining biodiversity and facilitate wildlife habitat (Naiman *et al.*, 1993; Schultz *et al.*, 2004).

A healthy river can fulfill the demand for drinking, agriculture, industry, fish, goods and services and other life supporting activities (Gregory *et al.*, 1991). The significance of the riparian zones to the aquatic ecosystem is well known, because the terrestrial primary productivity provides the important source of energy to riverine food-webs (Junk *et al.*, 1989; Medeiros *et al.*, 2006). The riparian vegetations help to retain nitrogen (Anbumozhi *et al.*, 2005; Dodds and Oakes, 2006) and phosphorus (Borin *et al.*, 2004) from upland diffuse and point sources of pollution (Meals, 2001; Sovik and Syversen, 2008). The riparian vegetation provides many critical functions in urban areas (Groffman *et al.*, 2003), pollution amelioration (Sweeney *et al.*, 2004), energy for stream organisms and temperature regulation (Johnson and Jones, 2000). They also offer unique habitat (Naiman and Decamps, 1997) and recreation opportunities. Diverse, native riparian assemblages, however, are disappearing from urban areas (Ozawa and Yeakley, 2007). The water quality is intimately connected to the riparian vegetal cover and hence, it can be used as key indicator and tool to evaluate the changes in water quality, biodiversity and river health (Sovik and Syversen, 2008).

Functions of Riparian Vegetation

Riparian vegetation have unique features and have great ecological and environmental functions. De Groot (1992), classified ecosystem functions into four general categories:

1. Regulation function,

2. Carrier or habitat function,

3. Production function, and

4. Information function.

The regulation utility associate with the capacity of natural and semi-natural ecosystems to govern constitutional ecological process and life support systems through bio-geochemical cycles and other biospheric processes. The habitat function or carrier function arises from the ability of an area to facilitate space and means to satisfy physical needs of humans, flora and fauna. The information functions arise from the provision of opportunities for enrichment, cognitive development and recreation afforded by natural ecosystems. The production functions are ecosystem goods that are produced naturally without alteration of natural process by mankind (De Groot, 1992).

The major key functions and services of riparian vegetation described as following:

a) **Biodiversity:** Riparian vegetation exhibit diversity in their structure, composition and ecological processes. The fluctuation in riparian topography, redistribution of sediment and high productivity potential of plant communities due to the proximity of water, collectively contribute towards creating a species rich environment than adjacent upland habitats within the same geographic location (Naiman *et al.*, 1993). The linear structure, regular floods, strength of competitive interactions, periodical successional stages and shifting mosaic of landforms provides a diversity of microhabitats, resulting high species richness and biodiversity in the riparian sites (Kalliola and Puhakka, 1988).

b) **Water quality and Flood Management:** Riparian vegetation intercepts and confine agricultural runoff from adjoining upland areas and wastewater pollution and maintains the water quality (Jones *et al.*, 1999). Riparian vegetation provides flood control during high rain events (Welsch *et al.*, 2000). It controls processes related to surface and subsurface flow at the local scale (Darby, 1999). The living vegetation forms point structures whereas dead plant debris forms mobile, resistant or labile structures and can obstruct, divert or facilitate water flow.

c) **Inputs for food-web:** Riparian vegetation facilited organic inputs addition as litter and coarse woody debris by senescence from riparian vegetation which are major food sources for various organisms in aquatic and terrestrial ecosystem and help to regulate and maintain food chains and food webs (Cummins, 1974).

d) **Biomass, Productivity and Temperature regulation:** The high productivity of riverine vegetation has been well-documented (Brinson *et al.*, 1981). The severe magnitude of floods may remove tree biomass (Stromberg, 1993), maintaining the forest in an early successional stage and creating the potential for positive net ecosystem production (Schade *et al.*, 2002). Biomass production differed significantly between the vegetation types, with a higher production in the forested sites (Hefting *et al.*, 2005). The canopy and stems of riparian vegetation above the ground provide shade, which controls temperature and in-stream photosynthetic productivity. These vegetation potentially moderates stream temperature and light levels, thereby influencing habitat favourable for fish and other aquatic organisms (Gregory *et al.*, 1991).

e) **Facilitate Corridors and Wildlife habitats:** The rich diversity of native vegetation found in riparian areas fosters high species richness and abundance of wildlife. Riparian vegetation functions as cover for wildlife and corridors for species migration, dispersal (Wegner and Merriam, 1979; Cordes *et al.*, 1997), breeding sites for birds, mammals and other wild animals (Blair, 1996; Cockle and Richardson, 2003). Riparian areas are particularly vital to wildlife, as they provide water, shelter and food necessary for survival. As the continuous rise of human population, large scale of riparian habitat are being altered or destroyed, making it increasingly difficult for riparian dependent wildlife to find sufficient space to live and other functions.

f) **Micro-habitats for fish and invertebrates:** The qualitative impact of coarse woody debris on stream hydraulics causes water flow diversions, congestions in the main channel, reduced connectivity and enhanced local erosion (Maser and Sedell, 1994). Riparian zone contributing large woody debris to streams and thereby shaping stream habitats such as pools and riffles and influencing sediment routing (Zimmerman *et al.*, 1967; Cummins 1974) and provide substrate for biological activity especially for fishes and other aquatic organisms (Nakamura and Swanson, 1994).

g) **Vegetational succession dynamics:** Riparian forests influenced both by disturbance agents of upland ecosystems and aquatic systems such as wind, fire, debris flows, lateral channel erosion and flooding. Among which, the hydrological processes interacts with the earliest stages of plant succession and has significant impact on pioneer vegetation. The patchiness of pioneer vegetation increases the heterogeneity of water flow patterns over sediment and vegetation mosaic, leading to the development of preferential flow pathways (Thorne *et al.*, 1997).

h) **Biological invasions and river regulation:** Riparian corridors are landscape elements that are most sensitive to plant invasion (DeFerrari and Naiman, 1994; Brown and Peet, 2003). The high frequency of open ground for colonization (Malanson, 1993), dispersal networks connecting different landscapes (Forman and Godron, 1986), frequent flooding and linear structure makes them highly susceptible to invasion by exotics (Planty-Tabacchi *et al.*, 1996). River regulation drastically alters the flood plain environment resulting reduction of species richness (Nilsson *et al.*, 1991) and increase of exotic species (Decamps *et al.*, 1995).

Riparian Vegetation Functional Mechanisms

Hickin (1984) was identify that riparian vegetation plays an important role in river system and detailed it as relating to the five mechanisms- through increasing flow resistance, through increasing bank strength, bar sedimentation, formation of log jams and concave bench deposition. Furthermore, Brooks and Brierley (2002) were identify three key mechanisms of vegetation that allow it to mediate the equilibrium channel condition as follow-

1. Mechanisms that physically resist change

2. Mechanisms reducing the impetus for change (lowering energy within the system)

3. Mechanisms enabling channel recovery following deviation from an equilibrium condition.

Floods play at least thrice-fold role in influencing the distribution of areas for vegetation establishment and the survival of riparian plants (Bendix and Hupp, 2000):

1. Most riparian plants germinate in alluvium that is deposited during floods

2. Floods may create colonisation sites by destroying pre-existing vegetation

3. The occurrence or lack of floods subsequent to germination may determine whether seedlings survive to maturity.

Riparian Vegetation and Soil Environment

The riparian vegetation and riparian soil are interdependent and mutually benefited to each other. The soil characteristic define the vegetation growth in the riparian areas and vegetation types and presence in the riparian area may improve and build-up suitable soil environment. Riparian and upland soils are derived from various parent material due chiefly to the influence of water on riparian soils. While the parent material of upland soils is generally the rock that underlies the site, the mineral component of riparian soils originates as stream-deposited sediment. Thus, riparian soils are potentially more heterogeneous in mineral character than their upland counterparts. Periodic sediment deposition to riparian areas by streams during floods is accompanied by the flushing of organic litter from riparian sites by water. This increases the heterogeneity of riparian soils by producing a bare soil surface in some areas. This phenomenon also increases plant diversity in riparian zones by creating hospitable micro-environments for the seeds of species that require a bare soil surface for germination (Bilby, 1988).

In addition to the redistribution of organic matter in riparian sites, the aquatic system further influences soil organic matter by increasing soil moisture. Riparian zones possess high levels of soil moisture due to both the presence of water in the stream and to the movement of groundwater into the rooting zone of riparian vegetation (Bilby, 1988). Fluctuations in groundwater levels that influence the nature and extent of plant production and microbial activity are responsible for much of the variation in riparian zone soil, plant and microbial properties.

In particular, impacts of livestock grazing on riparian soils include soil compaction, breakdown of undercut streambanks, and increased loss of sediment due to excessive removal of stabilizing vegetation. Timber harvest increases soil erosion and alters soil microclimates by increasing soil temperatures (Hall, 1988).

Stocking (1994) reported that vegetation is believed to be most suitable important factor for soil erosion control in tropics. Vegetation cover provides erosion protection to the soil by the interception of raindrops during rainfall and by absorbing their kinetic energy. Moreover, she also reported that if the vegetation is maintained above a certain level of cover, the interactive process between the soil and the plant are sufficient to limit erosive forces. There are a number of interactive processes between plant and soil that affect the degree of erosion including the physical binding of the soil by plant roots and stems, detention of runoff by plant stalks and organic litter and vegetation interaction with soil, all leading to better soil bonding and structure and improved infiltration (Stocking, 1994). The effects of vegetation on soil erosion control can be subdivided into three groups. First, height of the vegetative cover from surface is important in influencing droplets size. Second, the height of the canopy due to tall growing trees may reduce the level of ground cover completely. Last, there exists a complex interaction among vegetation, slope, soil type and erosion (Stocking, 1994).

Nutrient input, cycling and dynamics

Riparian vegetation significantly reduces nutrient runoff (Lowrance *et al.*, 1984) and helps in nutrient cycling (Vorosmarty *et al.*, 2000) and their role in nutrient dynamics (Cummins, 1992) were reported. Riparian forests serve as filters, transformers, sources and sinks for nutrients, sediment and pollutants associated with agriculture and urban runoff (Malanson, 1993).

Accelerated decomposition rates result in rapid nutrient release to riparian soils (Edmonds, 1980). The types and amounts of stream-deposited sediments also alter chemical composition of riparian soils. Types and amounts of vegetation present in riparian zones affect nutrient cycling in these areas. Certain elements like phosphorus and nitrogen tend to be rapidly taken up and conserved by riparian vegetation, while others, like Ca, Cl and Mg, show limited uptake by these plants (5-40% of total available) and are generally found in high concentrations in stream water (Bilby, 1988). Some nutrients, like nitrate, exhibit seasonal fluctuations in concentrations in streams, indicating that uptake by plants may be limited to the growing season.

Riparian vegetation has an effect on soil transfer processes in riparian zones. Riparian vegetation add organic material inputs which may form pools that reinforce channel structure and slow the water movement. Large woody debris and standing trees may create a barrier to the movement of sediment and litter material, increasing the height of streambank terraces and decreasing the likelihood of flooding in these areas. Presence of vegetation also exerts a strong influence on nutrient cycling in riparian zones.

Bank stability and Reducing agricultural runoff

The bank stability of river and adjoining part depends upon various factors including the presence of vegetal cover and vegetation mix. Roots of trees and other vegetation bind the soil together. Riparian vegetation stabilizing riverbanks by their root system (Cordes *et al.*, 1997), contributing root strength that maintains stream bank integrity (Broderson, 1973). Riparian forests function as buffers to reduce the quantity of agricultural diffuse pollution that reaches streams (Lowrance *et al.*, 1984). Nitrogen removal can be efficient in riparian zones (Hill, 1996; Mander *et al.*, 2000) by the denitrification process and prevents eutrophication (McClain *et al.*, 2003). The nutrient loads capture by riparian vegetation may vary according to vegetation types (table 1) and the potential and efficiency of pollutant removal may fluctuate as per the nature of pollutants (table 2). These riparian vegetation performs various function and pollutant removal efficiency was determined by vegetational stratum (table 3).

Table 1: Computation step-down of nutrient loads by riparian buffers or vegetation (Palace, 1998)

Vegetation/Buffer Types	Nitrogen	Phosphorus	Sediments
Vegetated filter strips	4-70%	24-85%	53-97%
Forested	48-74%	36-70%	70-90%
Forested and Vegetated Filter strips	75-95%	73-79%	92-96%

Table 2: Pollutants removal or retaintion by riparian vegetation from surface water runoff before to reach into freshwater (Gabor *et al.*, 2004).

Pollutants	Retention in Percentage
Sediments	66-97
Nitrogen	35-96
Phosphorus	27-97
Pesticides	8-100
Fecal colliform	70-74

Table 3: Removal efficiency by plant type (Fischer and Fischenich, 2000; Jontos, 2004)

Functions	Grass/Herbaceous layer	Shrub layer	Tree layer
Trapping sediment	High	Medium	Low
Filtration potential and efficiency (nutrients, insecticides, pesticides etc.)	High	Low	Low
Soluble form of nutrients and pesticides	Medium	Low	Medium
Flood conveyance	High	Low	Low
Bank stabilization and erosion control	Medium	High	High

Sedimentation and floodplain development

The term "sedimentation" encompasses three major processes of material transfer: erosion, transport and deposition (Swanson *et al.*, 1982). Sedimentation is an important process in riparian ecosystems, particularly for nutrient redistribution and export. Patterns of erosion and deposition create unique landforms that offer a variety of habitat opportunities for vegetation and microorganisms alike. Erosion also influences patterns of succession during or following disturbance on a local and regional level (Swanson *et al.*, 1982). Grassy riparian areas trap sediments delivered from hill slopes by overland flow (Magette *et al.*, 1989). The colonization of newly deposited sediments by dense herbaceous vegetation also helps to sustain high moisture levels in the upper sediment layers during dry periods because of the sheltering of the sediment surface by the vegetation cover and the capillarity provided by the rhizosphere.

Impacts and Importance of riparian areas in a nutshell

❖ Riparian areas offer greater biodiversity than the surrounding landscape (Hansen, 2000)

❖ Play a key role in maintaining high water quantity (Ehrhart and Hansen, 1997)

❖ Important for maintaining high water quality (Preston and Bedford, 1988)

❖ Facilitate valuable wildlife habitat and corridors (Cross, 1985)

❖ Provide a valuable forage source for livestock grazing (Fitch and Adams, 1998)

❖ Facilitate good habitat for fisheries (Wesche *et al.*, 1985)

❖ Offer high recreation value (Hoover *et al.*, 1985)

❖ Offer migration routes for plants and extending the range of plant species than normal limit (Coupland and Rowe, 1969).

Restoration Perspectives of Riparian Vegetation

Due to human civilization and pressure exerted on the natural resources alter the aquatic and terrestrial ecosystem in a great way. These changes may degrade the originality, quality and functioning which leads to unhealthy ecosystem and environment. To improve and maintain healthy riparian areas in the face of the increasing, and often conflicting, demand for multiple land use, they must be managed properly (Heady and Child, 1994). Riparian area management must be based on an understanding of the riparian ecosystem, its biotic and abiotic components as well as the interactions among those components and should employ the established principles of management (Fitch and Adams, 1998).

Riparian ecosystems, usually representing small parts of the landscape, are often the focus of intensive human activity, and present numerous challenges for managers because it can serves various roles for mankind (Kemper, 2001). Many types of human-mediated disturbances, occurring at scales from local to global, influence riparian ecosystems. A vast area of riparian vegetation has been degraded or removed where intensive agriculture under practices or due to transforming the land under agriculture land-use (Schultz *et al.*, 2004). From benefits perspective of riparian area, restoration or rehabilitation of degraded riparian zones will have a significant positive impact on the ecosystem and environment.

However, past history of disturbances or human impact has degraded these ecosystem severely and put it under the circumstances from where it become very difficult task to regain its originality to prior condition (Clewell et al., 2004). Therefore, restoration of these area required more practical approach which would be restoring ecosystem services such as carbon sequestration, soil erosion control, aggregate formation and stabilization, retention of sediments, nutrients and other contaminants. This would result in better water quality, conversion or reduction of incoming substances in aquatic ecosystem (Osborne and Kovacic, 1993).

The importance of the riparian zone and its functions has been widely recognized and therefore, restoring a permanent vegetative cover along stream banks is essential and encouraged (Schultz et al., 2004). In this regard, combination of fast growing trees, native bushes, grasses and shrubs which together function as a sink for nutrient and various pollutants runoff from agricultural fields were more effective measures (Schultz et al., 1995). The large root system of trees planted next to the stream is efficient in stabilizing the stream bank, shrubs provide diversity in both above and below ground plant structure and the rapid turnover of roots and large amounts of organic matter from grasses help to redevelop aggregate structure, soil microbial activity and soil microbial biomass thereby improving soil quality and health (Schultz et al., 2004). The soil condition and its health can be improved through reforestation activities and it reported that the riparian vegetation positively influenced both terrestrial and aquatic ecosystems (Schultz et al., 2004).

Moreover, the effective grazing management in riparian area is an important considerable points to recover or restore the natural sites in a sustainable way. The livestock grazing have markable effects on riparian characteristics such as stream channel morphology, hydrology and stream flow, water quality, riparian zone soils, and riparian and aquatic organism. Healthy riparian vegetation and soils are important for hydrological processes including flow-energy dissipation, bank building, ground water storage, aquifer recharge and flood control. Vegetation reduces stream velocity, thereby decreasing erosion and increasing deposition of materials on banks and shores. Therefore, proper protection and grazing management is essential consideration for proper implementation and riparian zone restoration and conservation implications.

Riparian zones fulfill important ecological roles for both aquatic and terrestrial ecosystems. Riparian restoration to mitigate damage to riparian ecosystems and to the streams and terrestrial landscapes with which they are integrally linked.

Overwhelming evidence shows that stock access in waterways leads to progressive and continued damage, whilst excluding stock halts decline and some passive recovery is possible.

Active restoration shows further potential in re-establishing the structure and composition of plant communities and capacity to recover broader ecological functions of riparian zones.

Riparian zones are among the most effective and efficient parts of the landscape to target for restoration and riparian restoration can help improve water quality, aquatic biodiversity, terrestrial biodiversity and the resistance and resilience of populations to stressors including climate change.

The capacity for restoration efforts to reinstate the ecological benefits of intact riparian zones depends on identifying and addressing the multiple drivers of degradation and potential constraints to recovery.

R & D plan and Future Perspectives

Past studies and scientific literature provide guidelines for the effective design and implementation of restoration efforts:

❖ Targets should be set to inform on ground works.

❖ Multiple drivers of degradation and potential constraints to recovery should be identified, prioritised and addressed.

❖ An adaptive monitoring regime should be employed to inform and improve restoration efficiency and effectiveness over time.

The design and delivery of the riparian extension program is highly recommended

❖ Riparian site description and classification, including carrying capacity and stocking rate recommendations, should be developed. Such information should be incorporated into the existing upland range site classification and guide.

❖ Improved and simplified methodology for riparian condition assessment and monitoring for use by producers is also required.

❖ Research is needed to address grazing management questions such as rest requirements for key riparian plants and stubble or residue requirements for key and indicator riparian plants.

❖ Scientific research is needed to evaluate the most promising riparian grazing management practices and approaches in relation to riparian health, with particular emphasis on water quality effects.

Assessments of riparian health are useful to managers because they generally reflect productivity potential, the effects of disturbances and potential response to future management. The riparian area health assessment described by Ehrhart and Hansen (1997) can be used to determine if vegetation, soils, and hydrology are representative of a riparian area that is properly functioning, functional at risk, or non functional. The classification is based on an assessment of: (1) vegetative cover of floodplain and stream banks, (2) percent cover of invasive species, (3) percent cover of disturbance-induced species, (4) amount of tree and shrub establishment and regeneration, (5) degree of tree and shrub utilization, (6) number of dead trees, (7) ability of roots to hold stream banks together, (8) proportion of human-caused bare ground in the riparian area, (9) human induced changes on stream bank structural, (10) amount of pugging and hummocking, and (11) the degree of channel incisement.

Conclusions

Riparian areas are complex, diverse and dynamic ecosystems. These are unique in its biological and physical characteristics as well as their past use history, making it difficult to draw generalized management recommendations. The dynamic nature of riparian areas represents a major challenge for the producer as it necessitates a great deal of skill, efforts and attention to properly monitor vegetation and other riparian health parameters. Rivers are conduits for materials and energy; this, the frequent and intense disturbances that these systems experience, and their narrow, linear nature, create problems for conservation of biodiversity and ecosystem functioning in the face of increasing human influence. A variety of factors contribute to the heterogeneity of soil and plant attributes in riparian zones. Some information is currently available about aboveground processes in riparian regions, such as plant succession, competition and response to disturbance, but very little information exists regarding riparian soils. Future research in riparian areas should continue to focus on below-ground processes like nutrient cycling and decomposition, as well as gaining new information about soil microbial biomass and soil microorganism populations and their role in soil processes along with grazing management and riparian area protection, conservation, restoration and management implication. A better understanding of these relationships can help to improve various ecological and environmental functions, restoration and maintenance of these ecosystems.

References

Agee, James K. (1988). Successional Dynamics in Forest Riparian Zones. p. 31-43. In: Raedeke, Kenneth J. (ed.) Streamside Management: Riparian Wildlife and Forestry Interactions. University of Washington. College of Forest Resources. Contribution Number 59.

Anbumozhi, V., Radhakrishnan, J. and Yamaji, E. (2005). Impact of riparian buffer zones on water quality. *Ecol. Engi.*, 24: 517–523.

Baker, T.T. (2002). What is a riparian area? Cooperative Extension Service Animal Resources Department. New Mexico State University. Las Cruces, NM. Available at: http://cahe.nmsu.edu/riparian/WHTRIPAREA.htm (Accessed on 08-12-05).

Blair, R.B. (1996). Land use and avian species diversity along an urban gradient. *Ecological Applications*, 6(2):506–519.

Brinson, M.M., Swift, B.L., Plantico, R.C. and Barclay, J.S. (1981). Riparian ecosystems: their ecology and status. US Fish and Wildlife Service, Kearneysville, West Virginia.

Brinson, M.M. (1990). Riverine forests. In: Lugo, A.E., Brinson, M.M. and Brown, S. (Eds) *Ecosystems of the World - 15: Forested wetlands*. Elsevier, Oxford. pp. 87–141.

Brooks, A.P. and Brierley, G.J. (2002). Mediated equilibrium: the influence of riparian vegetation and wood on the long-term evolution and behaviour of a near-pristine river. *Earth Surface Processes and Landforms.* 27:343-367.

Broderson, J.M. (1973). *Sizing buffer strips to maintain water quality. M.Sc. Thesis,* University of Washington, Seattle, Washington. 86 pp.

Brown, R.L. and Peet R.K. (2003). Diversity and invasibility of Southern Appalachian plant communities. *Ecology,* 84:32–39.

Bendix, J. and Hupp, C.R. (2000). Hydrological and geomorphological impacts on riparian plant communities. *Hydrological Processes.* 14: 2977-2990.

Bilby, Robert E. (1988). Interactions between Aquatic and Terrestrial Systems. p. 13-29. In Raedeke, Kenneth J. (ed.) Streamside Management: Riparian Wildlife and Forestry Interactions. University of Washington. College of Forest Resources. Contribution Number 59.

Borin, M., Bigon, E., Zanin, G. and Fava, L. (2004). Performance of a narrow buffer strip in abating agricultural pollutants in the shallow subsurface water flux. *Environ. Pollu.* 131: 313–321.

Clewell, A., Aronson, J. and Winterhalder, K. (2004). The SER international primer on ecological restoration. Society for Ecological Restoration International Science, Tucson, Arizona.

Cockle, K.L. and Richardson, J.S. (2003). Do riparian buffer strips mitigate the impacts of clear cutting on small mammals? *Biological Conservation,* 113:113–140.

Connell, J.H. (1978). Diversity in tropical rain forests and coral reefs. *Science,* 199:1302–1310.

Cordes, L.D., Hughes, F.M.R. and Getty, M. (1997). Factors affecting the regeneration and distribution of riparian woodlands along a northern Prairie River: The Red Deer River, Alberta, Canadia. *Journal of Biogeography,* 24:675–695.

Coupland, R.T. and Rowe, J.S. (1969). Natural vegetation of Saskatchewan. P73-78. In: Atlas of Saskatchewan, University of Saskatchewan, Saskatoon, SK.

Cross, S.P. (1985). Responses of small mammals to forest riparian perturbations. *In*: Riparian Ecosystems and Their Management: Reconciling Conflicting Uses. First North American Riparian Conference. April, 1985. Tucsun, AZ. 523 pp.

Cummins, K.W. (1974). Structure and function of stream ecosystems. *BioScience,* 24:631–641.

Cummins, K.W. (1992). Catchment characteristics and river ecosystems. In: Boon P.J., Calow, P. & Petts, G.E. (Eds) *River conservation and management.* John Wiley & Sons, Chichester, UK, pp. 125–135.

Darby, S. (1999). Effect of riparian vegetation on flow resistance and flood potential. *Journal of Hydrological Engineering,* 125(5):443–453.

Decamps, H., Planty-Tabacchi, A.M. and Tabacchi, E. (1995). Changes in the hydrological regime and invasions by plant species along riparian systems of the Adour River, France. *Regulated Rivers: Research and Management,* 11:23–33.

DeFerrari, C.M. and Naiman, R.J. (1994). A multi-scale assessment of the occurrence of exotic plants on the Olympic Peninsula, Washington. *Journal of Vegetation Science,* 5:247–258.

De Groot, Rudolf S. (1992). *Functions of Nature. : Evaluation of Nature in Environmental Planning, Management and Decision Making*.xviii + 315p, illustrated, soft cover. ISBN 90-01-35594-3. Dfl 80.

Diamond, J. (2005). The world as a polder: What does it all mean to us today? p. 486-499. *In* The world as a polder: What does it all mean to us today? Collapse: How societies choose to fail or succeed. Viking Press, New York.

Dodds, W.K. and Oakes, R.M. (2006). Controls on Nutrients Across a Prairie Stream Watershed: Land Use and Riparian Cover Effects. *Environmental Management,* 37:634–646.

Dudgeon, D. (1992). Endangered ecosystems: A review of the conservation status of tropical Asian rivers. *Hydrobiologia,* 248:167–191.

Dudgeon, D. (2000). The ecology of tropical Asian rivers and streams in relation to biodiversity conservation. *Annual Review of Ecology and Systematics,* 31:239–263.

Dudgeon, D.,Arthington, A. H. and Gessner, M.O. (2006). Freshwater biodiversity: importance, threats, status and conservation challenges. *Biological Reviews,* 81: 163–182.

Edmonds, R.L. (1980). Litter decomposition and nutrient release in Douglas-fir, red alder, western hemlock, and Pacific silver-fir ecosystems in western Washington. *Can. J. For. Res.,* 10:327-337.

Ehrhart, R.C. and Hansen, P.L. (1997). Effective cattle management in riparian zones: a field survey and literature review. United States Department of Interior, Bureau of Land Management, Montana State Office. Missoula, Montana. Riparian Technical Bulletin No. 3. 92 pp.

FAO. (2010). Global Forest Resources Assessment. Main Report. Available from http://www.fao.org/docrep/013/i1757e/i1757e.pdf.

Fischer, R.A. and Fischenich, J.C. (2000). Design recommendations for riparian corridors and vegetated buffer strips. U.S. Army Engineer Research and Development Center, Environmental Laboratory. Vicksburg, MS.

Fitch, L. and Adams, B.W. (1998). Can cows and fish co-exist? *Can. J. Plant Sci.* 78:191-198.

Foley, J.A., DeFries, R., Asner, G.P., Barford, C., Bonan, G., Carpenter, S.R., Chapin, F.S., Coe, M.T., Daily, G.C., Gibbs, H.K., Helkowski, J.H., Holloway, T., Howard, E.A., Kucharik, C.J., Monfreda, C., Patz, J.A., Prentice, I.C.,

Ramankutty, N. and Snyder, P.K. (2005). Global consequences of land use. *Science*, 309:570-574.

Forman, R.T.T. and Godron, M. (1986). Landscape Ecology. Wiley, New York.

Gabor, T.S., North, A.K., Ross, L.C.M., Murkin, H.R., Anderson, J.S. and Raven, M. (2004). *Natural Values-The Importance of Wetland and Upland Conservation Practice in Watershed Management: Function and Values for Water Quality and Quantity.* Ducks Unlimited Canada unpublished report. 55 pp.

Gadgil, M. (2004). *Karnataka state of environment report and action plan biodiversity sector.* ENVIS Technical Report No.16. Indian Institute of Science, Bangalore.

Goebel, P.C., Hix, D.M., Dygert, C.E. and Holmes, K.L. (2003). *Ground flora communities of headwater riparian areas in an old-growth central hardwood forest.* USDA Forest Service General Technical Report NC-234. pp. 136–145.

Gopal, B. (1988). Wetlands: management and conservation in India. *Water Quality Bulletin*, 13: 3-6.

Gregory, S.V., Swanson, F.J., McKee, W.A. and Cummins, K.W. (1991). An Ecosystem Perspective of Riparian Zones. *Bioscience*, 41:540-551.

Groffman, P.M., Bain, D.J., Band, L.E., Belt, K.T., Brush, G.S., Grove, J.M., Pouyat, R.V., Yesilonis, I.C. and Zipperer, W.C. (2003). Down by the riverside: Urban riparian ecology. *Frontiers in Ecology and the Environment*, 6: 315–321.

Hall, Frederick C. (1988). Characterization of Riparian Systems. p. 7-12. In Raedeke, Kenneth J. (ed.) Streamside Management: Riparian Wildlife and Forestry Interactions. University of Washington. College of Forest Resources. Contribution Number 59.

Hansen, P. (2000). Riparian systems and their functions. In: The Range: progress and potential. Western Range Science Seminar. Jan. 23-25, Lethbridge, Alberta, Alberta Agriculture, Food and Rural Development and Agriculure and Agri-Food Canada, Lethbridge, Alberta.

Hefting, M.M., Jean-Christophe Clement, Bienkowski, P., Dowrick, D., Guenat, C., Butturini, A., Topa, S., Pinay, G. and Verhoeven, J.T.A. (2005). The role of vegetation and litter in the nitrogen dynamics of riparian buffer zones in Europe. *Ecological Engineering*, 24:465–482.

Heady, H.F. and Child, R.D. (1994). Rangeland ecology and management. Westview Press. San Francisco, CA. 519 pp.

Hickin, E.J. (1984). Vegetation and river channel dynamics. *Canadian Geographer* 28(2):111-126.

Hill, A.R. (1996). Nitrate removal in stream riparian zones. *J. Environ. Qual.*, 25:743-755.

Hoover, S.L., King, D.A. and Matter, W.J. (1985). A wilderness riparian environment: visitor satisfaction, perceptions, reality, and management. *In*: Riparian Ecosystems and Their Management: Reconciling Conflicting Uses. First North American Riparian Conference. April, 1985. Tucson, Arizona. 523 pp.

Hynes, H.B.N. (1970). *The ecology of running waters*. Liverpool University Press, UK. 555 pp.

Johnsingh, A.J.T. and Joshua, J. (1989). The threatened gallery forest of the River Tambiraparani, Mundanthurai Wildlife Sanctuary, South India. *Biological Conservation*, 47:273–280.

Johnson, S.L. and Jones, J.A. (2000). Stream temperature responses to forest harvest and debris flows in western Cascades, Oregon. *Canadian Journal of Fisheries and Aquatic Sciences*, 57: 30–39.

Jones, E., Helfman, G., Harper, J. and Bolstad, P. (1999). Effects of riparian forest removal on fish assemblages in Southern Appalachian streams. *Conservation Biology*, 13:1454–1465.

Jontos, R. (2004). Vegetative buffers for water quality protection: an introduction and guidance document. Connecticut Association of Wetland Scientists White Paper on Vegetative Buffers. Draft version 1.0. 22pp.

Junk, W.J., Bayley, P.B. and Sparks, R.E. (1989). *Aqu. Sci.* 106: 110-127.

Kalliola, R. and Puhakka, M. (1988). River dynamics and vegetation mosaicism: a case study of the River Kamajohka, northernmost Finland. *Journal of Biogeography*, 15:703–719.

Kemper, N.P. (2001). *RVI: Riparian vegetation index*. WRC Report 850/3/01. Water Research Commission, Pretoria, South Africa.

Leonard, S.G., Staidl, G.J., Gebhardt, K.A. and Prichard, D.E. (1992). Viewpoint: Range site/ecological site information requirements for classification of riverine riparian ecosystems. *J. Range Manage.*, 45:431-435.

Lowrance, R., Todd, R., Fail, J., Hendrickson, Jr. O., Leonard, Jr. R. and Asmussen L. (1984). Riparian forests as nutrient filters in agricultural watersheds. *Bioscience*, 34:374-377.

Magette, W.L., Brinsfield, R.B., Palmer, R.E. and Wood, J.D. (1989). Nutrient and sediment removal by vegetated filter strips. *Transactions of the American Society of Agricultural Engineers*, 32:663–667.

Malanson, G.P. (1993). Riparian landscapes. Cambridge University Press, Cambridge, UK, 296 p.

Manci, Karen M. (1989). Riparian Ecosystem Creation and Restoration: A Literature Summary. Biol. Rep. 89(20) Fish and Wildlife Serv., U.S. Dep. of the Interior. Washington D.C.

Mander, U., Kull, A., Kuusemets, V. and Tamm, T. (2000). Nutrient runoff dynamics in a rural catchment: Influence of land-use changes, climatic fluctuations and ecotechnological measures. *Ecological Engineering*, 14: 405–417.

Maser, C. and Sedell, J.R. (1994). *From forest to the sea: The ecology of wood in streams, rivers, estuaries and oceans*. St. Lucie Press, USA.

McClain, M.E, Boyer, E.W., Dent, C.L., Gergel, S.E., Grimm, N.B., Groffman, P.M., Hart, S.C., Harvey, J.W., Johnston, C.A., Mayorga, E., McDowell, W.H. and

Pinay, G. (2003). Biogeochemical hot spots and hot moments at the interface of terrestrial and aquatic ecosystems. *Ecosystems,* 6:301–312.

Meals, D.W. (2001).Water quality response to riparian restoration in an agricultural watershed in Vermont, USA. *Water Sci. Technology.,* 43: 175–182.

Medeiros, E.S.F., Ramos, R.T.C., Ramos, T.P.A. and Silva, M.J. (2006). Spatial variation in reservoir fish assemblages along a semi-arid intermittent river, Curimataú River, northeastern Brazil. Rev. Biol. Ciênc. Terra. *Supl. Esp.* 1: 29-39.

Minore, D. and Weatherly, H.G. (1994). Riparian trees, shrubs and forest regeneration in the coastal mountains of Oregon. *New Forests,* 8: 249–263.

Naiman, R.J., Decamps, H. and Pollock, M. (1993). The role of riparian corridors in maintaining regional biodiversity. *Ecol. Appl.,* 3:209-212.

Naiman, R.J. and Decamps, H. (1997). The ecology of interfaces: Riparian zones. *Annual Review of Ecology and Systematics,* 28: 621–658.

Nakamura, F. and Swanson, F.J. (1994). Distribution of coarse woody debris in a mountain stream, Western Cascade Range, Oregon. *Canadian Journal of Forest Research,* 24:2395–2403.

Nilsson, C., Ekblad, A., Gardfjell, M. and Carlberg, B. (1991). Long-term effects of river regulation on river margin vegetation. *Journal of Applied Ecology,* 28:963–987.

Ortega Klett, C.T. (2002). New Mexico Water Rights. 2nd Update. New Mexico Water Resources Research Institute. New Mexico State. Las Cruces, NM.

Osborne, L.L. and Kovacic, D.A. (1993). Riparian vegetated buffer strips in water-quality restoration and stream management. *Freshwat. Biol.,* 29:243-258.

Ozawa, C.P. and Yeakley, J.A. (2007). Performance of management strategies in the protection of riparian vegetation in three Oregon cities. *Journal of Environmental Planning and Management,* 50: 803–822.

Palace, M., Hannawald, J., Linker, L., Shenk, G., Storrick, J., Clipper, M., 1998. Chesapeake Bay Watershed Model Application and Calculation of Nutrient and Sediment Loadings Appendix h: Tracking Best Management Practice Nutrient Reductions in the Chesapeake Bay Program. Chesapeake Bay Program Office, Annapolis, MD. EPA 903-R-98–009, CBP/TRS 201/98.

Pettit, N.E. and Froend, R.H. (2001). Variability in flood disturbance and the impact on riparian tree recruitment in two contrasting river systems. *Wetland Ecology and Management,* 9: 13–25.

Planty-Tabacchi, A.M., Tabacchi, E., Naiman, R.J., DeFerrari, C. and Decamps, H. (1996). Invasibility of species rich communities in riparian zones. *Conservation Biology,* 10:598–607.

Preston, E.M. and Bedford, B.L. (1988). Evaluating cumulative effects on wetland functions: a conceptual overview and generic framework. *Environmental Management,* 12:565-583.

Schade, J.D., Marti, E., Welter, J.R., Fisher, S.G. and Grimm, N.B. (2002). Sources of nitrogen to the riparian zone of a desert stream: Implications for riparian vegetation and nitrogen retention. *Ecosystems*, 5:68–79.

Schultz, R.C., Isenhart, T.M., Simpkins, W.W. and Colletti, J.P. (2004). Riparian forest buffers in agroecosystems - lessons learned from the Bear Creek watershed, Central Iowa, USA. *Agrofor. Syst.*, 61-2:35-50.

Schultz, R.C., Colletti, J.P., Isenhart, T.M., Simpkins, W.W., Mize, C.W. and Thompson, M.L. (1995). Design and placement of a multispecies riparian buffer strip system. *Agrofor. Syst.*, 29:201-226.

Sedell, James R., Robert J. Steedman, Henry A. Regier and Stanley V. Gregory. (1991). Restoration of Human Impacted Land-Water Ecotones. p. 110-129 In Holland, Marjorie M., Paul G. Risser, and Robert J. Naiman (eds.) Ecotones: The Role of Landscape Boundaries in the Management and Restoration of Changing Environments. Chapman and Hall. New York, NY.

Smol, J.P. et al. (2005). Climate-driven regime shifts in the biological communities of arctic lakes. Proceedings of the National Academy of Sciences of the United States of America 102: 4397–4402.

Sovik, A.K. and Syversen, N. (2008). Retention of particles and nutrients in the root zone of a vegetative buffer zone - effect of vegetation and season. *Boreal Environ. Rese.*, 13:223-230.

Stocking, M.A. (1994). Assessing vegetative cover and management effects. In: Lal, R. ed., *Soil Erosion Research Methods*, 2nd edition, Soil and Water Conservation Society, Ankeny, IA, 211-232.

Stromberg, J.C. (1993). Instream flow models for mixed deciduous riparian vegetation within a semiarid region. *Regulated Rivers*, 8:225–235.

Swanson, F.J., Fredriksen, R.L. and McCorison, F.M. (1982). Material Transfer in a Western Oregon Watershed. p.233-266. In Robert L. Edmonds (ed.) Analysis of Coniferous Forest Ecosystems in the Western United States. Hutchinson Ross Pub. Stroudsburg, PA. 1982.

Sweeney, B.W., Bott, T.L., Kaplan, L.A., Newbold, J.D., Standley, L.J., Hession, W.C., Horwitz, R.J. and Wolman, M.G. (2004). Riparian deforestation, stream narrowing, and loss of stream ecosystem services. *Proceedings of the National Academy of Sciences of the USA*, 101: 14132–14137.

Tabacchi, E., Planty-Tabacchi, A.M. and Décamps, O. (1990). Continuity and discontinuity of the riparian vegetation along a fluvial corridor. *Landscape Ecology*, 5: 9–20.

Thorne, C.R., Amarasinghe, I., Gardiner, J. and Sellin, R.P.G. (1997). *Bank protection using vegetation with species reference to willows*. Engineering and Physical Sciences Research Council and Environmental Agency Report, UK.

Tockner, K., Ward, J.V., Edwards, P.J. and Kollmann, J. (2002). Riverine Landscapes: an introduction. *Freshwat. Biol.* 47(4): 497-500.

UNEP. (2014). Climate change mitigation. Available from http://www.unep.org/climatechange/mitigation/Introduction/tabid/29397/Default.aspx (accessed February 19, 2014).

Vannote, R.L., Minshall, G.W., Cummins, K.W., Sedell, J.R. and Cushing C.E. (1980). The river continuum concept. *Canad. J. Fishe. Aqu. Sci.* 37:130-137.

Vorosmarty, C.J., Gessner, M.O., Dudgeon, D., Prusevich, A., Green, P., Glidden, S., Bunn, S.E., Sullivan, C.A., Reidy Liermann, C. and Davies, P.M. (2010). Global threats to human water security and river biodiversity. *Nature*, 467: 555–561.

Vorosmarty, C.J., Fekete, B.M., Meybeck, M. and Lammers, R.B. (2000). Global system of rivers: its role in organizing continental land mass and defining land to ocean linkages. *Global Biogeochemical Cycles*, 14: 599–621.

Wegner, J. and Merriam, G. (1979). Movement by birds and small mammals between a wood and adjoining habitats. *Journal of Applied Ecology*, 16:349–357.

Welsch, D.J., Hornbeck, J.W., Verry, E.S., Dolloff, C.A. and Greis, J.G. (2000). Riparian management: Themes and recommendations. In: Verry, E.S., Hornbeck, J.W. & Dolloff, C.A. (Eds) *Riparian management in forests of the Continental Eastern United States*. Lewis Publishers, Boca Raton, FL.

Wesche, T.A., Goertler, C.M. and Frye, C.B. (1985). Importance of instream and riparian cover in smaller trout streams. *In*: Riparian Ecosystems and Their Management: Reconciling Conflicting Uses. First North American Riparian Conference. April, 1985. Tucson, Arizona. 523 pp.

Zimmerman, R.C., Goodlett, J.C. and Comer, G.H. (1967). The influence of vegetation on channel form of small streams. *International Association of Scientific Hydrology Publication*, 75:255-275.

Impact of Textile Effluents on Soil Fertility in Pali District of Rajasthan

Meenu Srivastava and Vinita Koka

Department of Textiles & Apparel Designing, College of Home Science, MPUAT, Udaipur

Abstract

Present study throws light on impact of textile effluents on soil fertility. Textile processing units of Pali is of small scale and labour intensive industry. The major problem threatening the textile processing units is the environmental pollution arising out of wet processing of textiles. Textile effluent on being discharged is also absorbed by soil by making it unhealthy and unfit for growing vegetables and crops. In order to assess the quality of soil, samples ,were collected from the farms of eight selected villages adjoining the Bandi River were analyzed on different parameters.

Keywords: Impact, textile, effluents, soil, fertility, Pali, Rajasthan

Introduction

The textile industry of India also contributes nearly 14 per cent of the total industrial production of the country. It also contributes around 3 percent to the GDP of the country. (http://business.mapsofindia. com/india-industry/textile.html) Rajasthan has leading position in spinning of polyester, viscose yarn and synthetic suiting and processing, printing and dyeing of low cost, low weight fabric at Pali, Balotra, Sanganer and Bagru.(http://www.rajasthantour4u.com/business/ textile.html). Now a days, mechanical process and chemical dyes are used, which generate polluted effluents. At places viz. Pali, Balotra, Jasol, Bithuja, Jodhpur, Sanganer and Bagru, there are concentrations of large number of small scale units of textile dyeing and printing, which discharge water containing dyes and other chemical pollutants. (http://rajasthantextile.com/aboutrajasthan.html)

Environmental pollution and occupational health hazard is the severe problem due to rapid industrialization. The phenomenon is very common in textile industries. Textile industry need large quantity of water for different processes and their demand is increasing every year with expansion of industries. Apart from consuming large quantity of water, industry also discharge a variety of effluents. Disposal of industrial waste is the major problem responsible for soil and water pollution. Pollution is the main accuse in the textile processing units.

The effluent discharged by these industries lead to serious pollution of surface water and ground water. Singh and Bhati (2007) conducted a study to assess the impact of effluents on the changes in soil properties and their effect on soil nutrient availability (N, P, K, Ca, Mg, Na, Cu, Fe, Mn and Zn), and on seedling nutrient status, survival and growth on six-month-old seedlings of *Acacia nilotica*, *Dalbergia sissoo* and *Eucalyptus camaldulensis* and found that textile effluents increased Na, and decreased Mg and micronutrient concentration in both soil and seedlings.

Pali is one of the polluted areas identified by the Central Pollution Control Board in 1998. The industries discharge a variety of chemicals, dyes, acids and alkalies besides heavy metals and other toxic compounds. Textile effluents discharged from various textile processing units of Pali, flow about 55 Kilometer downstream, making the ground water in several riverbank villages unfit for drinking and irrigation and also causes adverse effect on crops productivity and health of people residing in those areas. Before disposal they need to be treated for certain acceptable tolerance limits since pollution control laws are strictly followed all over the world and captured worldwide attention. The use of toxic chemicals in these units cause threat to the manpower employed in such units in a way directly resulting in occupational health hazards. Further to be in tune with the government restrictions to be connected to CETP, majority of textile processing houses/units of Pali district are now adjoined to CETP. Inspite of the installation of CETP, the Bandi River still have enormous water and soil pollution adversely affecting the fertility of soil and purity of drinking water.

In view of the above facts, and the researcher's past experiences of being a local resident of Pali city, such problems are personally observed in those affected areas hence the researcher got the idea to conduct the present investigation with the objectives to study the profile of textile processing units of Pali district of Rajasthan and to assess the impact of textile effluent on environment in terms of soil contamination in selected areas.

Methods of investigation

The present study was conducted in Pali district of Rajasthan as it has the largest number of textile processing units in the Bandi basin. A list of textile processing units was procured from District Industries Centre (DIC) Pali, which was registered since last 20 years. Out of these, 30 units each among cotton textile processing and synthetic textile processing units connected with Common Effluent Treatment Plant was purposively identified and selected, whose effluent was directly or indirectly being discharged in Bandi river in order to assess the impact of effluent on environment in terms of soil and water contamination after being treated with CETP. A pre structured interview schedule was developed to gain information from the respondents i.e. Head (Manager/Director) of the textile processing units regarding the profile of both cotton as well as synthetic units.

In order to assess the quality of soil, samples were collected from the farms of eight selected villages adjoining the Bandi River. These were analyzed on different parameters such as pH, EC, Available Nitrogen, Available phosphorous, Available potash, organic carbon and heavy metals like Zn, Cu, Fe and Mn. All the parameters mentioned above for analysis of quality of soil samples were assessed by using standard test methods.

Major findings

Profile of Textile Processing Units: The first industrial area in Pali, established in 1962. At present 867 textiles industrial units are in operation. All the units are in operation at small scale due to lack of capital. There are about four RIICO developed industrial area in the city. Mandia Road industrial area- phase III alone has about 525 small scale textile units, which is 60.55 percent of the total dyeing and printing industries. Majority of textile processing units were selected from Mandia Road Industrial area due to higher concentration in that area.

The salient feature of textile processing units of the study area is that different textile processes related to production are done by separate units. The number of composite units doing both dyeing and printing are only 70 i.e. 8.07 percent of total units.

It was found that majority (66.66%) of textile processing units were established between year 1990-2000 followed by 25 percent sample units in year 1980-1990. Majority (35%) of textile processing units covered the area between 2000-3000 square feet while 26 percent textile processing units were spread between 3000-4000 square feet area. Total capital investment was found between 3 to 4 crore by 36.66 percent followed 30 percent between 2 to 3 crore.

Data on production per month was found in between 10 to 15 lac meters; that means average production of approximately 300 'than' for dyeing and 150 'than' for printing per day. One 'than' is equal to near about 130 meters. 40 percent of responding units had yearly turnover between 20-30 crore, 31.66 percent between 30-40 crore and only 10 percent had 40-50 crore.

Raw Materials

The raw materials used in dyeing and printing industry of the study area included the grey cloth (loom cloth), various chemicals and dyes. Almost all the raw materials were brought from the industrial units situated outside of the basin. The main centers were Ahmedabad, Malegaon, Mumbai, Bhiwandi, Ichalkaranji, Surat and Tripur. A small quantity of cloth was also brought from the power loom of Kishangarh in Rajasthan.

All the respondents used synthetic dyes for dyeing and printing. Very few respondents (16.66%) also used eco-friendly dyes, for fabric coloration as per the export order procured. Mainly two types of fabrics were used for dyeing and printing i.e. cotton and polyester. For cotton, mainly Reactive, Ramazol, High Exhaustive dyes, Azoic dyes, vat dyes and sulphur dyes was used. Disperse dye was applied on most of the synthetic fabric. For polyester, it is only for colouring.

Procion and pigment dyes were used for printing of cotton fabric and disperse dyes for polyester printing.

Requirement of water

Adequate water supply is the biggest problem in textile dyeing and printing units of Pali. The reservoirs of Hemavas and wells along the bank of Bandi River supply the requirement of water to the inhabitants and also to the dyeing printing units. Water requirement per day in textile processing units depends on the size of the unit and production capacity. Approximately 5 to 6 liter of water on an average is needed up to the final product of one meter cloth. Majority of the units (46.66%) required 1.5 lac to 2 lac liter water per day for textile processing followed by 25 units whose water requirement is 2 to 2.5 lac liter per day.

Wastewater discharged

The waste water generation by the dyeing and printing units is due to various activities preformed on grey clothes to shape the final products. All the respondents reported that 60 to 70 percent of total water consumed in textile processing is being discharged as 'waste water' after final finishing of the products.

Man Power Requirement

Dyeing and printing industry being a part of textile industry is labour intensive. Since labour cost accounts for up to 60 percent of total cost in dyeing and printing, it provides an important location influences. The availability of plentiful supply of cheap labour is an important factor in the establishment of textile processing industries in the study area. Number of employed workers in different units depends on the load of work or production capacity of particular textile processing unit. 45 percent of units had 25 to 50 workers for its production. Only 6.66 percent units had more than 100 workers.

Majority of textile processing units (81.66%) runs in both shifts. It means industry runs 24 hours a day continuously. Remaining textile processing units had production in one shift only further in textile processing units majority of workers were found unskilled (60%) or semiskilled (30%) based on their experience. Skilled labour (10%) required for mixing of dyes, printing of clothes and dyeing of polyester fabric.

Market

One of the prominent features of textile processing units of the study area is that 45 per cent of units have their sister concerns in the big towns and cities of Gujarat, Maharashtra, Karnataka, Tamilnadu and West Bengal. These concerns are also wholesale dealers of the products of the textile units. Therefore through these dealers, the developed products capture the market all over the India. Major export countries are Saudi Arabia, Emirates, Pakistan, Myanmar, Bangladesh and Srilanka.

Transport and production cost

Sourcing of raw materials and chemicals used for production is carried out from distant places. The transport cost of the clothes for the purpose of dyeing and printing is kept Rs. 0.30 per meter. The variance in this cost corresponds to the distance from the source of raw material. The production cost includes the cost incurred on input to produce one unit of cloth. Generally the dyeing cost is kept at the rate of Rs. 4.50 to Rs. 5.00 per meter. The rate of printing varied from Rs. 6.50 to Rs. 7.00 per meter. The variation in rate depends on the quality of dyes, printing and size of the clothes.

Quality parameters

Majority of respondents (70%) did not have testing lab for quality control. Only 16.66 percent respondents follow quality standards as per requirement with export orders. 83.33percent respondents did not follow any quality standards because they produce cheap low quality fabric for lower and lower middle class people at low cost.

Subunits in textile processing

Each textile processing unit had different subunits. These subunits performed different functions required for the fabric to be finished. First four processes are called preparatory processes which prepare the fabric for dyeing and printing. These textile processing used various non-ecofriendly chemicals in different processes which caused hazards to environment and health of workers.

Impact on Environment and Health of Worker

Majority of the respondents were agreed that textile processing units created hazardous effect on the environment through water and soil pollution (86.66%). Working environment and chemicals that are being used in various processes created hazardous effect on the health of the workers as reported by 76.66 percent respondents. All the respondents not provided safety features at workplace. About 40 per cent units had first aid and fire extinguisher (56.66%) facility as safety features to meet out some accidents at the time of working.

No effluent treatment

Majority of units had no provision of primary treatment being of small scale industry. It was evident that respondents did not reuse the waste water; they directly discharged it in the drains which were connected to common effluent treatment plant. Very few respondents (23.33%) used to give primary treatment to the waste water to maintain pH. Hence most of contaminated water used to deposit in the bed of Bandi River flowing adjacent to town thus causing water, air and soil pollution

Physiochemical analysis of soil samples

Textile effluent on being discharged is also absorbed by soil by making it unhealthy and unfit for growing vegetables and crops. In order to assess the quality of soil, samples ,were collected from the farms of eight selected villages adjoining the Bandi River were analyzed on different parameters such as pH, EC, Available Nitrogen, Available phosphorous, Available potash, organic carbon and heavy metals like Zn, Cu, Fe and Mn.

pH –Soil pH is one of the most indicative measurements of chemical properties of soil. It indicates acidic, neutral or alkaline nature of soil. The pH values in the study area vary from 7.82 to 10.18 suggesting the alkaline nature of soil. All the soil samples except S3, S4 and S6 were found above the permissible limit of 8.5.

Electrical Conductivity, (EC)- Table clearly indicates that EC in soil samples vary from 1 to 1.82 μ mho/cm^2. All the soil sample except S3 showed higher EC rating as compared to standard (>1μ mho/cm^2). Hence these samples are highly unsuitable and injurious for proper growth of plants.

Available Nitrogen- Available Nitrogen is the amount of Nitrogen present as either nitrate or ammonium forms which can be readily take up by plant.

Subbiah and Asija (1956) have established the following rating for the classification of soil test values of available nitrogen.

Low : < 250 kg/ha

Medium : 250-500 kg/ha

High : > 500 kg/ha

On the basis of these limits, the available Nitrogen was found in low range in all the soil sample, which indicates the need of using nitrogen based fertilizers.

Available phosphorous: The following rating has been suggested for classifying the available phosphorus content into low, medium and high categories (Muhr *et al.*, 1965).

Low	:	Below 20 kg P_2O_5/ ha
Medium	:	20 to 50 kg P_2O_5/ Ha
High	:	Above 50 kg. P_2O_5/ ha

On the basis of these limits, sample S1, S3, S4, S6, and S7 fall in low range of available phosphorous and samples S2, S5 and S8 comes under medium range, which also suggest application of phosphate fertilizer to maintain desired level of phosphorous in the soil..

Organic carbon: Organic carbon serves as an efficient index of availability of nitrogen to plants. Therefore, organic carbon has been used for evaluating fertility status of soil. The range of organic carbon in the soil is 0.5 to 0.75 per cent. In the study area, organic carbon varied between 0.12 to 0.44 percent. It shows that the soil is unfertile for production due to which, the plants can not absorb all the available nutrients. Hence, addition of organic manure or FYM (Farm Yard Manure) is necessary to all the soil samples to improve soil fertility.

Table 1 : Physio-Chemical analysis of soil samples collected from selected villages in and around Bandi River, Pali

S. No.	Soil Parameters	Unit	Standard	S1	S2	S3	S4	S5	S6	S7	S8
1	pH		6.5-7.5	9.31	10.18	8.24	8.67	9.72	7.82	9.25	10.33
2	Ec at 25°C	μ mho/ cm^2	>1	1.03	1	0.34	1.25	1.42	1.07	1.33	1.82
3	Nitrogen	kg/ha	250-00	185	190	175	160	177	184	179	186
4	Phosphorous	Kg/ha	23-56	8.3	26.55	15.07	18.22	29.52	12.56	18.77	21.24
5	Organic carbon	%	0.5-0.75	0.44	0.19	0.41	0.39	0.29	0.36	0.42	0.22
6	Potassium	kg/ha	150-300	743.9	372.0	206.6	425.32	625.21	815.13	723.70	620.24
7	Zinc (Zn)	mg/kg	0.6-1.2	0.458	0.442	0.548	0.432	0.562	0.444	0.437	0.456
8	Copper (Cu)	mg/kg	0.2	0.618	0.922	0.876	0.726	0.928	0.8621	1.726	0.989
9	Iron (Fe)	mg/kg	4.5	4.821	2.780	4.714	4.358	4.628	4.812	4.989	4.52
10	Manganese (Mn)	mg/kg	1.0	10.83	5.88	6.23	7.48	8.62	7.22	6.48	8.81

Available Potash : The following rating is being used for classifying the available phosphorus content into low, medium and high categories. (Muhr *et. al.*, 1965).

Low   :   <125 kg K_2o/ha

Medium   :   125-300 kg K_2o / ha

High   :   > 300 kg K_2o/ha

On the basis of these limits, all the soil samples except S4 were falling in the high range of available potassium.

Heavy metals: The table further depicts the content of different heavy metals in selected soil samples Takkar and Mann (1975) suggested limits of available Zn as <0.6, 0.6 and >.12 mg/kg and 1.2 mg/kg of DTPA extractable in for deficient, marginal and sufficient classes respectively. Considering these limits soil sample, S1, S3, S7, S8 were found deficient in Zn supply, sample S2, S4, S6 were found marginal in Zn supply. Lindasay and Norvell (1978) Suggested 0.2mg/kg of DTPA extractable Cu is critical limit of variable copper. Considering these limits all the soils were found to be high in available copper and 4.5 mg/ha of DTPA extractable iron (Fe) as critical limit of available iron. Considering these critical limits all sample except S1, S7, and S8 found sufficient in available iron. Considering 1.0 mg/kg of DTPA extractable Mn is critical limit for available manganese, all soil samples were found vary high in manganese content.

Nickel is a transition metal and found in natural soils at trace concentrations Nickel was not detected (ND) in all the soil samples.

Thus it can be concluded that soils are quite alkaline in nature due to high pH. EC is also found high than the standard value. Organic carbon, N, P, K were also not found in satisfactory range. Hence application of specific fertilizers and manures are required for improving soil fertility. Heavy metal like, Cu, Mn and were found higher than permissible limits which causes hindrance in plant growth.

Physiochemical analysis of soil showed that the soils are quite alkaline due to high pH. Electrical conductivity is also found high than the standard value, is directly related to total dissolved solids. Organic carbon, potash, and phosphorus were also not found in satisfactory range. Soil is becoming more and more unproductive due to irrigate with well water or dam water. Therefore these soils are suffering from high salinity, which is not favorable for growth of plants. The structure of the soil has changed and become fragile and porous with salt crystal and seeds did not germinate after irrigation with polluted water.

Conclusion

Hence it can be concluded that textile processing units of Pali is of small scale and labour intensive and operating with the limitations. The major problem threatening the textile processing units is the environmental pollution arising out of wet processing of textiles. Huge amount of water and chemicals are used in different processes are discharged as waste water that are high in COD, BOD, TDS and toxic chemicals. Inspite of the installation of CETP, the Bandi River still have enormous water and soil pollution adversely affecting the soil fertility, purity of drinking water and health of people residing in nearby villages around the Bandi River. Findings clearly reveals that the disposal of contaminated water without proper treatment takes place on the ground which acts as sand bed. The sandy soil percolates almost all the contents of waste water to the ground water. Such percolation imparts alkalinity to the soil and makes it unfertile. Excessive dissolved salts are also creating osmotic pressure and therefore preventing growth of plants which require more energy for using water for its growth.

Physiochemical analysis of soil and water of study area revealed that contamination of ground water and soil by toxic chemicals has become wide spread and is posing serious health and environmental problems. To minimize the impact of future incidents by adopting clean technologies and zero discharge system in textile processing units. Many soil remediation techniques i.e. reclamation of contaminated soil, bioremediation techniques etc. can be used to improve soil fertility.

References

Lindsay, W. L. and Norvell, W. A. (1978). Development of a DTPA test for Zinc, Iron, Manganese and Copper. *Journal of American Society of Soil Science*, **42:** 421-428.

Muhr, G.R., Datt, N.P., Sankasuramoney, H., Ieiey, V.K., Donahue, L. and Roy (1965). Soil Testing in India, United States Agency for International Development, New Delhi, pp.20.

Pandey, J and Sharma, M.S. 2000 Environment Sciences Practical and field manual Yash publishing House, Bikaner.

Singh, G. and Bhati, Madhulika. 2007. Effect of mixed industrial effluent on soil properties and survival of trees seedlings. *The Indian society of soil science*. www.cababstractplus.org.

Subbiah, B.V. and G.L. Asija, 1956. A rapid procedure for the estimation of available nitrogen in soils. Curr. Sci., **25:** 259-260.

Takkar, P.N. and Mann, M.S. (1975). Evaluation of analytical methods for estimating available zinc and response of maize to applied zinc in major soil series of Ludhiana, Punjab (India). Agrochemica, 19 (5) : 420-430.

(http://business.mapsofindia. com/india-industry/textile.html)

(http://www.rajasthantour4u.com/business/ textile.html)

(http://rajasthantextile.com/aboutrajasthan.html)

An Introduction to Soil Contamination

Preeti Sharma, Geetanjly, AnkitaVerma and Anchala Nautiyal

College of Agriculture, G.B. Pant University of Agriculture and Technology, Pantnagar-263145 (Uttarakhand) India.

Abstract

Soil Contamination is becoming a major threat to the agriculture, especially as populations and industrial economies expand. With an ever-increasing population and expanding industrial economies, there is a corresponding increase in environmental pollution. Contaminants in roadside soil can easily be found in many areas of the world, and these contaminants, some of which are metals, can be identified and quantified. Some metals, such as lead and manganese, are not biologically useful. Even if metal has biological functions, they can exist in too high of concentrations so as to be toxic, such as the case with iron, zinc and copper. This toxicity leaves plants struggling to live in such polluted soils. To decrease the detrimental impact of this pollution, some changes to the environment can be made. Phytoremediation, or the rehabilitation of soils by use of plants, can be easily utilized.

Keywords:

Definition

Soil contamination is defined as the build-up in soils of persistent toxic compounds, chemicals, salts,radioactive materials, or disease causing agents, which have adverse effects on plant growth and animal health (Okrent, 1999).

The European Commission has proposed the following definition of 'contaminated site': a site where there is a confirmed presence, caused by human activities, of hazardous substances to such a degree that they pose a significant risk to human health or the environment, taking into account land use (Commission Proposal COM (2006) 232).

Introduction

Soil is the basic need in Agriculture. All crops for food production and animal feed depend upon it. Soils are formed over many years due to decomposition of rock and organic matter. Soil properties vary with differences in its composition, climate, and other factors. If variations occur in the level of some soil elements and other substances fit can affect health of humans, animals, or plants. Soil properties may affect by various factors as past land use, current activities on the site, and site location. Human activities contaminate the soil by various means such as pesticides, fertilizers and other amendments to soils. Accidental spills and leaks of chemicals used for commercial or industrial purposes and some contaminants are moved through the air and deposited as dust have also been sources of contamination. There are various factors for regulation of pollutants in the eco system among them soil is major one which plays a fundamental role. Soil is the interface for most human activity and is greatly impacted by humans (Mohan and Sajayan, 2015). Soil Pollution is as dangerous as pollution of water and air. Rural areas are also not freed from soil pollution since they are using various fertilizers and pesticides for agriculture. Plants retain those toxins and when they die, it decompose the toxic material back into the soil which lead to the residue problem and enter into food chain in due course the soil will become unusable. Soil bears the greatest burden of environmental pollution. There is urgency in controlling the soil contamination in order to preserve the soil fertility and increase the productivity (Ashraf *et al.,* 2014). Assessing the ecological risk of contaminated soil leading to exposure of the terrestrial environment to hazardous substances is a complicated task with numerous associated problems. In order to preserve the fertility and the productivity of the soil, control measures are to be taken in a herculean manner, thereby improving the health of all living beings. Government has taken various criteria's and policies to access the level of nutrients and control over use of inorganic chemicals and disposal of non-bio degradable waste. The measures are mainly Sustainable taken with the prospective of increasing the soil quality for increased productivity of crops. Since soil health may directly affect the living beings, more measures are needed regarding health. According to the National Bureau of Soil Survey and Land Use Planning (NBSS&LUP, 2004) ~146.8Mha is degraded. Based on first approximation analysis of existing soil loss data, the average soil erosion rate was ~16.4 ton ha−1year−1, resulting in an annual total soil loss of 5.3 billion tons throughout the country (Dhruvanarayan and Ram, 1983). Nearly 29% of total eroded soil is permanently lost to the sea, while 61% is simply transferred from one place to another and the remaining 10% is deposited in reservoirs (Srinivasarao *et al.,* 2013).

Table 1. Land degradation in India, as assessed by different organizations.

S. No.	Organizations	Assessment Year	Degraded Area (Mha)
1	National Commission on Agriculture	1976	148.1
2	Ministry of Agriculture-Soil and Water Conservation Division	1978	175.0
3	Department of Environment	1980	95.0
4	National Wasteland Development Board	1985	123.0
5	Society for Promotion of Wastelands Development	1984	129.6

S. No.	Organizations	Assessment Year	Degraded Area (Mha)
6	National Remote Sensing Agency	1985	53.3
7	Ministry of Agriculture	1985	173.6
8	Ministry of Agriculture	1994	107.4
9	NBSS&LUP	1994	187.7
10	NBSS&LUP (revised)	2004	146.8

(NBSS & LUP, 2004)

Types of soil contamination

◈ Agricultural Soil Pollution

◈ Soil pollution by industrial effluents and solid wastes

◈ Pollution due to urban activities

Sources of soil contamination

The main sources which pollute the soil are:

◈ Agricultural sources

◈ Non-agricultural sources.

a) Agricultural sources as contaminated soil, hazardous herbicide and pesticide application, sewage sludge amendment, and other human activities leading to exposure of the terrestrial environment to hazardous substances.

b) Non-agricultural sources are the direct result of urbanization caused by rapidly increasing population and a rapidly per capita output of waste materials related to our modem way of life. Its materials that find their entry into the soil system have long persistence and accumulate in toxic concentration and thus become sources of contamination(Swartjes, 1999).

FORMS OF SOIL CONTAMINANTS

Industrialization is better option for development but on the other side it is polluting our environment too as inorganic residues which contain metals in industrial waste cause serious problems. Industrial waste also emits large amounts of arsenic fluorides and sulphur dioxide ($SO2$) (Richardson *et al.*, 2006). Sulphur dioxide may make soils very acidic which can be harmful for crops. These metals cause leaf injury and destroy vegetation. Elements as Copper, mercury, cadmium, lead, nickel, and arsenic can accumulate in the soil. Some of the fungicides containing copper and mercury also add to soil pollution. Smokes from automobiles contain lead gets adsorbed by soil particles which can be toxic to the plants (Van Zorge, 1996).

1. Organic wastes

Organic wastes of various types contain borates, phosphates, detergents, phenols and coal in large amounts. Domestic garbage, municipal sewage and industrial wastes if improperly disposed, affect human beings, plants and animals. The main organic contaminants are Asbestos, combustible materials, gases like methane, carbon dioxide, hydrogen sulphide, carbon monoxide, sulphur dioxide, petrol are also contaminants. The radioactive materials like uranium, thorium, strontium etc. also cause dangerous soil contamination(Nathanail *et al.*, 2015).

a. Sewage and sewage sludge

Sewage and other liquid wastes as industrial wastes, agricultural effluents from animal husbandry and drainage of irrigation water and urban runoff often contaminate the Soil (Tarazona, *et al.*, 2005). Irrigation with sewage water causes profound changes in the irrigated soils as physical changes like leaching, changes in humus content, and porosity etc., chemical changes like soil reaction, Base Exchange status, salinity, quantity and availability of nutrients like nitrogen, potash, phosphorus, etc. Accumulating of metals like lead, nickel, zinc, cadmium also occurs may lead to the phytotoxicity of plants (Evans, 2006).

b. Heavy metal pollutants

Heavy metals are widely distributed in the environment, soils, plants, animals and in their tissues. These are essential for plants and animals in trace amounts. Urban and industrial aerosols, combustion of fuels, liquid and solid from animals and human beings, mining wastes, industrial and agricultural chemicals etc. are the main components contributing heavy metal pollution. Heavy metals are also present in soils which can contaminate our crops too. Applications of chemicals, sewage sludge, farm slurries, etc. Increased doses of fertilizers, pesticides or agricultural chemicals, over a period, add heavy metals to soils which may contaminate them. Certain fertilizers contain phosphate and cadmium which may accumulate in the soils (Urzelai *et al.*, 2000).

2. Organic pesticides

Farmers use pesticides in several ways to control insect-pest problem, but due to their hazardous use for crops create many problem to our environment. Among which residue problem is major one. Pesticides which are not rapidly decomposed may create such problems. Accumulation is residues of pesticides in higher concentrations are toxic. Pesticides persistence in soil and movement into water streams may also lead to their entry into foods and create health hazards. Pesticides and its transformation products could be grouped into :(a) Hydrophobic, persistent, and bio-accumulable pesticides. The pesticides are retained by soils to different degrees, depending on the interactions between soil and pesticide properties. Organic matter content, the greater the adsorption of pesticides. Pesticides particularly aromatic organic compounds is not degraded rapidly and therefore, has a long persistence time. Mercury, cadmium and arsenic are common constituents of pesticides and all these heavy metals are toxic (Apitz, 2008).

Causes of soil pollution

Urbanization, Industrialization and Mining

Industrialization, urbanization and infrastructure development is taking away considerable areas of land from agriculture, forestry, grassland and pasture, and unused lands with wild vegetation. Negative effects of mining are water scarcity due to lowering of water table, soil contamination, removal from mine area results in significant loss of vegetation and rich topsoil (Sahu and Dash, 2011).

Natural and Social Sources of land Degradation

Natural causes of land degradation include earthquakes, tsunamis, droughts, avalanches, landslides, volcanic eruptions, floods, tornadoes, and wildfires. Some underlying social causes of soil degradation are land shortage, decline in per capita land availability, economic pressure on land, land tenancy, poverty, and population increase.

Use of fertilizers, fungicides, insecticides and herbicides indiscriminately

Main cause of Soil contamination is alteration in the natural soil environment. Industrialization and regular pesticide uses are contaminating soil environment regularly. This type of contamination arises from the rupture of underground storage links, application of pesticides, percolation of contaminated surface water, oil and fuel dumping, leaching of wastes from landfills or direct discharge of industrial wastes to the soil. Among them major chemicals are petroleum hydrocarbons, solvents, pesticides, lead and other heavy metals. Soil quality is important for plant growth and development in which nutrients play an important role. Plants obtain necessary nutrients like nitrogen, phosphorus, potassium,calcium, magnesium, sulfur and more must be obtained from the soil. Fertilizers are beneficial for plant growth as they use for nutritional deficiency in soil. Fertilizers impurities come from the raw materials used for their manufacture can contaminate soil. Fertilizers have metal content since the metals are not degradable. Due to excessive use of phosphate fertilizers,their accumulation in the soil above their toxic levels becomes poison for crops. The over use of NPK fertilizers reduce quantity of vegetables and crops grown on soil over the years. It also reduces the protein content of wheat, maize, grams, etc., grown on that soil (Huinink, 1998). Excess potassium fertilizers in soil decreases Vitamin C and carotene content in vegetables and fruits and more prone to attacks by insects and disease.

Food is basic requirement for living beings. Insect pest and disease harm our crops at different crop growth stages. To kill unwanted populations of them farmers use pesticides/ fungicides on their crops. It is acceptable that these chemicals are beneficial but very harmful if use hazardously. Use of DDT in earlier time is the best example for residue problem and still we are facing its harmful effects. Insect became resistant to DDT and as the chemical did not decompose easily, it persists in the environment. The pesticides used on pests may get adsorbed by the soil particles, which then contaminate root crops grown in that soil. The consumption of such crops causes the pesticides remnants to enter human biological systems, affecting them adversely. Pesticides not only bring toxic effect on human and animals but also decrease the fertility of the soil. Some of the pesticides are quite stable and their bio-degradation may take weeks and even months. Pesticide problems such as resistance, resurgence, and health effects have caused scientists to seek alternatives.

However, the regular use of fungicides can potentially pose a risk to the environment, particularly if residues persist in the soil or migrate off-site (e.g. due to spray drift, run-off). If this occurs it could lead to adverse impacts to the health of terrestrial and aquatic ecosystems. Long term use of copper-based fungicides, which can result in an accumulation of copper in the soil. This in turn can have adverse effects on soil organisms (e.g. earthworms, microorganisms) and potentially pose a risk to the long-term fertility of the soil (Provoost *et al.*, 2006).

Herbicides have widely variable toxicity. The acute toxicity due to exposure had led to long-term problems. The pathway of attack can be transported via surface run off and leaching to contaminate distant water source. Herbicides decompose rapidly in soil via soil microbes, hence inhibiting the activities of microorganism in the soil. Herbicides have caused significant drop in bird population by decreasing vegetation which bird rely and had linked to decline in seed eating species. Most of the chemicals used;in pesticide are persisted soil determinant which adversely affect soil conservation (Reddy, 2003).

Dumping of solid wastes

Commercial, industrial and agricultural operations produce solid waste includes garbage, domestic refuse and discarded solid materials as paper, cardboards, plastics, glass, old construction material, packaging material and toxic or otherwise hazardous substances. The majority of waste is recyclable or biodegradable in landfills and mining waste is left on site. The portion of solid waste that is hazardous such as oils, battery metals, heavy metals from smelting industries and organic solvents are the ones we have to pay particular attention to. These canin the long run, get deposited to the soils of the surrounding area and pollute them by altering their chemical and biological properties(Patterson *et al.*, 2007). Toxic chemicals leached from oozing storage drums into the soil underneath homes, causing an unusually large number of birth defects, cancers and respiratory, nervous and kidney diseases.

Deforestation

Soil erosion is the result of deforestation, agricultural development, temperature extremes, precipitation including acid rain, and human activities. Humans use various natural resources for their use as construction, mining, cutting of timber, over cropping and overgrazing. It results in floods and cause soil erosion. Forests and grasslands are an excellent binding material that keeps the soil intact and healthy which support our environment and ecosystems as feeding pathways or food chains to all species. Their loss would threaten food chains and the survival of many species. Due to this disturbance which we are creating to our environment the past few years quite a lot of vast green land has been converted into deserts. We have to pay attention to all these factors to save our environment (Leon, 2008).

Pollution of surface soils

Large quantities of city wastes including several Biodegradable materials and many non-biodegradable materials are cause of several problems such as:

◈ Clogging and leakage of drainage lines leading to health problems.

◈ Solid wastes have seriously damaged the normal movement of water, water accumulation cause various disease problem.

◈ Microbial decomposition of organic wastes generate large quantities of methane besides many chemicals to pollute the soil

◈ Hospital wastes as solid waste may create many health problems.

Pesticide contamination in soil: case studies

Residue levels of organochlorine, organophosphate pesticides and herbicides in agricultural soils from Delhi region has been studied by (Kumar et al., 2011). Among OCPs, HCH, DDT, endo-sulphan and dieldrin ranged between<0.01-104.14 ng g-1, <0.01-15.79 ng g-1, <0.01-7.57 ng g-1 and <0.01-2.38 ng g-1, respectively. The concentration of OPPs ranged from <0.01-20.95 ng g-1, ND-3.92 ng g-1, ND-31.73 ng g-1, ND-6.46 ng g-1and ND-6.46 ng g-1 for phosphomidon, monocrotophos, chlorpyriphos, quinolphos and ethion, respectively. Pendimethalin (0.27 ng g-1) was the dominant herbicides followed by but achlor (0.19 ng g-1), and fluchloralin (0.05 ng g-1). Their data showed the region was contaminated by technical DDT and technical HCH mixture. In this study the authors concluded that the level of some organochlorine pesticides in agricultural soils is a matter of concern for future food chain accumulation and human health so, regular investigation of pesticide residues is recommended on soil health and contamination levels. (Muruganet al., 2013) has studied DDT and its various metabolites, α-endosulfan, β-endosulfan, endosulfan sulfate, aldrin, and fenvalerate, in this area using the modified QuEChERS method for multi residue extraction from soils and detection with a gas chromatograph. Among the different pesticide groups detected, endosulfan and DDT accounted for 41.7 % each followed by aldrin (16.7 %) and synthetic pyrethroid (8.3 %). They investigated that a significantly higher concentration of pesticide residues in rice-vegetable grown in the valley followed by rice-fallow and vegetable-fallow in the coastal plains. They found that the soil microbial biomass carbon is negatively correlated with the total pesticide residues in soils, and it varied from 181.2 to 350.6 mg kg(-1). This study provides evidence that pesticide residues have adversely affected the soil microbial populations, more significantly the bacterial population. The Azotobacter population has decreased to the extent of 51.8 % while actinomycetes were the least affected though accounted for 32 % when compared to the soils with no residue.(Rao *et al.,* 2015) studied the pesticide residues during 2006–2009 in various crops and natural resources (soil and water) in the study village (Kothapally, Telangana State (TS)) and indicated the presence of a wide range of insecticidal residues. It was found that among 80 food crop and cotton samples, two rice grain samples (3 %) showed beta endosulfan residues, and for soil samples, two (3 %) soil samples showed alpha and beta endosulfan residues. In vegetables of the 75 tomato samples, 26 (35 %) were found contaminated with residues of which 4 % had residues above MRLs. Among the 80brinjal samples, 46 (56 %) had residues, of these 4 % samples had residues above MRLs. Only 13 soil samples from vegetable fields were found contaminated. They investigated that the frequency of contamination in brinjal fields was high and none of the pulses and cotton samples revealed any pesticide contamination. IPM fields showed substantial reduction sprays which in-turn reflected in lower residues. Yadav *et al.* (2015) reviewed the health impacts of persistent organic pesticides and their effect on environment. Pesticide traces can be detected in all areas of the environment (air, water and soil). They discussed the production and consumption of persistent organic pesticides, their maximum residual limit (MRL) and the presence of persistent organic pesticides in multicomponent environmental samples (air, water and soil) from India. According to their review, India is one of the major contributors of global persistent organic pesticides and they highlighted the impact of these persistent organic pesticides on neighboring countries. Indian soils of agricultural importance were found to be contaminated by some of the organochlorine pesticides. The DDT and HCH compounds were found as residues in rice, pulse, tea leaf, chili, pepper and cashew nut. Their review revealed that highest contamination of DDT and HCH was detected for tea leaf and pepper at Chennai and Chidambaram. An interesting fact indicated that human fat samples from Delhi and Ahmedabed were detected with higher accumulation of DDT and HCH, when compared with other states like Agra, Kolkata and Mumbai. In this study they concluded that, we must give the awareness programme to the Indian farmers about the hazards of organochlorine pesticides in living beings and our surrounding environment. (Fosu Mensah *et. al.,* 2016)studied the organochlorine pesticide residues in soils and water samples in cocoa growing areas in Ghana in 2016. They detected four organochlorine pesticide residues in soil samples: Indane (0.005–0.05 mg/kg), beta-HCH (<0.01–0.05 mg/kg), dieldrin (0.005–0.02 mg/kg), and p,p'-DDT (0.005–0.04 mg/kg), with dieldrin occurring most frequently. They concluded that these banned organochlorine pesticide residues in soil samples are still being used, illegally, on some cocoa farms. To minimize health risks, prevention, control and reduction of environmental pollution, routine monitoring of pesticide residues in the study area is necessary.

Causes in brief:

◈ Intensive industrial activities, inadequate waste disposal, mining

◈ Handling spills or accidents or insignificant losses/emissions.

◈ Municipal waste disposal, energy production and transport,

◈ Legacy of inefficient technologies and uncontrolled emissions

◈ Abandoned industrial facilities and storage sites

◈ Chemical fertilizer runoff from farms and crops

◈ Acid rain Sewage discharged into rivers instead of being treated properly

◈ Over application of pesticides and fertilizers

◈ Purposeful injection into groundwater as a disposal method

◈ Interconnections between aquifers during drilling (poor technique)

◈ Septic tank seepage

◈ Sanitary/hazardous landfill seepage

◈ Leaks from sanitary sewers

Soil contamination process

After enter into the soil, contaminants go through various process. Some carbon-based contaminants can undergo chemical changes or degrade into products that may be more or less toxic than the original compound. Chemical elements such as metals cannot break down, but their characteristics may change so that easily taken up by plants or animals. Site management and land use can affect some soil characteristics and helpful for soil management. Some contaminants end up in water held in the soil or in the underlying groundwater, volatilize into the air or bind tightly to the soil, vary in their tendency too. Important soil characteristics affect the behavior of contaminants as: Soil mineralogy and clay content (soil texture); pH (acidity) of the soil; Amount of organic matter in the soil; Moisture levels; Temperature; and Presence of other chemicals. The bioavailable portion is the amount of a substance cause direct effects on plants, animals or humans because it can be taken up by their bodies. All contaminant found in soil is not biologically available. The bioavailability of a contaminant depends on many characteristics of the soil and of the site. Site conditions affect how tightly the contaminant is held by soil particles and its solubility (how much of it will dissolve in water). More of the contaminant is bioavailable if Greater solubility but it also means that the contaminant is more likely to leach out of the soil. Some chemicals show an "aging effect "and bioavailable the longer they remain in soils. The bioavailable portion may be only a small fraction of the total amount, site conditions, such as soil acidity or organic matter content, can vary the bioavailability of a contaminant. Using bioassay tests to measure uptake of contaminants by plants or soil organisms is the most direct way to estimate bioavailability but bioassay tests are slow and expensive and are not generally available. For this reason, only the total levels or chemically extractable amounts of a particular contaminant are usually measured.

Chemicals may be carried by winds and deposited on the surface of soils; tilling can then mix these surface deposits into the soil. Groundwater or surface water may also affect how contaminants spread from the source (Rombke, 2006).Many pesticides and soil amendments used for agricultural, industrial, or commercial activities may be found in residential soils. Spills, runoff, or aerial deposition of chemicals used for agriculture or industry can also result in contamination of the soils of residential sites.

Effects of soil contamination

Agricultural

◈ Reduced soil fertility, nitrogen fixation,crop yield

◈ Deposition of silt in tanks and reservoirs

◈ Imbalance in soil fauna and flora, soil structure and nutrients

Industrial

◈ Dangerous chemicals entering underground water

◈ Ecological imbalance and reduced vegetation

◈ Release of pollutant gases and radioactive rays

◈ Increased salinity

Urbanization

◈ Clogging of drains, inundation

◈ Public health problems

◈ Pollution of drinking water sources

◈ Waste management problems

Environmental Long Term Effects of Soil Contamination

Contaminated soil should no longer be used to grow food, because the chemicals, toxic compounds or microbes present in the soil can leech into the plants, which is very much harmful to living beings. The land will usually produce lower yields. This, in turn, can cause even more harm because a lack of plants on the soil will cause more erosion, spreading the contaminants onto land. In addition, the pollutants will change the composition of the soil and the types of microbes present in it. In this way if whole food chain will be affected. If certain organisms die off in the area, the larger predator animals will also have to move away or die because they've lost their food supply. Thus it's possible for soil pollution to change whole ecosystems and contaminate the environment too.

Control of soil contamination

To secure our earth and environment pollution control is very much essential. In general we would need less fertilizer and fewer pesticides if we could all adopt the three R's: Reduce, Reuse, and Recycle. By using natural resources sustainable we can protect our ecosystem from contamination. The following steps have been suggested to control soil contamination:

Inter-cropping

Agronomical practices like use of cover crops, mixed/inter/strip cropping, crop rotation, green Manuring and mulch farming are vital practices associated with integrated nutrient management.

Growing soybean (*Glycine max*)/groundnut (*Arachishypogoea*)/cowpea (*Vignaradiata*) with maize (*Zea mays*)/jowar (*Sorghum bicolor*)/bajra (*Pennisetumglaucum*) is a common example of inter-cropping in the drylands. Wheat and mustard (*Brassica junceaL.*) crops ensured optimum use of space and soil moisture, increased wheat equivalent yield by 14% and net returns by 30% compared to mixed sowing (Singh *et al.*, 1993 and Sharma *et al.*, 2013).

Nutrient Management and Organic Manuring

Integrated nutrient management, *i.e.*, the application of NPK mineral fertilizers along with organic manure, increases crop productivity, and decreases soil contamination. Annual farmyard manure addition improved labile (movable; short-lived) and long-lived C pools (Bhattacharyya, *et al.*, 2011).

Desalinization

Liming is the most desirable practice for amelioration of acid soils. Lime raises soil pH, thereby increasing the availability of plant nutrients and reducing toxicity of Fe and Al. Tillage, irrigation and leaching reduce saline soil. Inversion tillage can decrease potential soluble salt accumulation in the root zone compared to zero tillage (Bhat *et al.*, 2010). Gypsum is the major chemical used for reclamation of alkali soils and phosphogypsum or acid formers like pyrites, sulphuric acid, aluminium sulphate and Sulphur are also used.

Remediation

Phytoremediation employing hyper-accumulating plants like brake fern (*Pterisvittata*) and water hyacinth (*Eichorniacrassipes*). Blue-green algae also have ability to decontaminate as of paddy soils through accumulation in its biomass and subsequent removal. These type of plants are eco-friendly and helpful in decreasing contamination (Ghosh *et al.*, 2004).

Water Management

Domestic and municipal wastes, sludges, pesticides, industrial wastes, *etc.* need to be used with utmost caution to avoid the possibility of soil. (Ambast and Sen, 2006) developed a user-friendly software 'RAINSIM' primarily for small holdings in the Sundarbans region based on hydrological processes, as well as in different agro-climatic regions. The software may be used for (i) computation of soil water balance; (ii) optimal design of water storage in the "on-farm reservoir" concept for converting up to 20% of the watershed; (iii) design of surface drainage in deep waterlogged areas to decrease water congestion in 75% of the area; and (iv) design of a simple linear program to propose optimal land allocation.

Intensive and Diversified Cropping and Integrated Farming Systems

There is already a greater emphasis on crop diversification due to growing concerns about the unsustainability. As example: the water requirement for rice is about 80% greater than for other crops. Growing non-rice crops in some areas and summer cropping with legumes such as green gram, cowpea (*Vigna unguiculata* L.) or dhaincha (*Sesbania sp.*) are essential for conserving resources and improving productivity.

Reducing chemical fertilizer and pesticide use

Sustainable use and IPM practices in proper way are the best alternative to reduce hazardous pesticide use. Eco-friendly approaches, bio-pesticides and applying bio-fertilizers and manures can reduce chemical fertilizer and pesticide use. Biological methods of pest control can also reduce the use of pesticides and thereby minimize soil contamination.

Reusing of materials

Waste materials such as glass containers, plastic bags, paper, cloth etc. can be reused at domestic level rather than being disposed, reducing solid waste pollution.

Recycling and recovery

Materials such as paper, some kinds of plastics and glass can and are being recycled. This decreases the volume of refuse and helps in the conservation of natural resources. For example, recovery of one ton of paper can save 17 trees.

Reforesting

Restoring forest and grass cover to check wastelands, soil erosion and floods are helpful in controlling land loss and soil erosion. Crop rotation or mixed cropping can improve the fertility of the land and also helpful to reduce soil contamination.

Solid waste treatment

For solid waste treatment Proper methods need to be adopted. Industrial wastes can be treated by various methods physically, chemically and biologically until they are less hazardous. Acidic and alkaline wastes should be first neutralized; the insoluble material if biodegradable should be allowed to degrade under controlled conditions as environmental and aesthetic considerations must be taken into consideration before selecting the dumping sites. Burying the waste in locations situated away from residential areas is the simplest and most widely used technique of solid waste management. Pyrolysis is a process of combustion in absence of oxygen or the material burnt under controlled atmosphere of oxygen, obtained gas and liquid can be used as fuels. Waste products as firewood, coconut, palm waste, corn combs, cashew shell, rice husk paddy straw and saw dust, yields charcoal along with products like tar, methyl alcohol, acetic acid, acetone and a fuel gas can be utilize in this process.

Natural land pollution:

Earth quakes, landslides, hurricanes and floods contaminate spoil in large amount. All cause hard to clean mess and may sometimes take years to restore the affected area. For recovery various initiatives should be done. As example., Disaster (Tsunami) Management following aspects are important to protect the soil (i) traditional disaster detection systems should be integrated with current scientific techniques; (ii) early warning systems need to be installed in coastal regions; (iii) protection against tsunamis can be achieved through construction of sea walls, beach defenses, tree plantations, and making buffer zones like raised land masses and forests; (iv) awareness about tsunamis and their impact in coastal areas has to be created not only among the public but also among officials; and (v) enforcement of by-laws and 'Coastal Regulation Zone Norms' should be strictly implemented to minimize tsunami damage (Bhattacharyya, *et al.*, 2015).A tsunami early warning system for the Indian Ocean was installed. The Indian Ocean Tsunami Warning System was agreed to in a United Nations conference.

Conclusion

Soil contamination can be a major risk if not controlled. Various contaminants after entering into the soil will enter into food chain and may affect the whole ecosystem, through which whole Agriculture Scenario will suffer. Sustainable use of natural resources and take need full action time to time can be a better option to save our environment as Sustainable agricultural approach using innovative farming practices have tremendous potential of increasing productivity and conserving natural resources, particularly by improving soil quality. Conservation agriculture (CA) is having various other technologies like micro-irrigation, fertigation, and management of problem soils using specific and necessary technologies hold great promise to increase productivity of crops and fruits and reverse soil degradation. Domestic and municipal wastes, sludge, pesticides, industrial wastes, *etc.* need to be used if possible to close nutrient cycles, but with caution to avoid the possibility of soil pollution. Future research should focus on enhancing soil nutrient and reduction in the pesticide residue problem. For promotion of these practices across diverse agro-ecologies, appropriate policy and institutional and technology support would be necessary. Local communities at every stage in the implementation of resource conserving technologies involvement are needed. Judicious irrigation water management, wasteland reclamation, watershed development, and afforestation are the major practices should be utilized for reducing soil contamination. An integrated land use policy, reforestation, to control fodder grazing and controlling fragmentation of land holdings is needed for sustainable management of land and forest. There is a need to increase awareness among people that prevention of adverse health effects and promotion of health are profitable investments for us and for sustainable development of economics. This could be achieved by land rights security and land tenure and encouraging the efficient and effective use of land. We are already seeing the longer-term trends and impacts of our industrial heritage and previous activities. It is important to understand why a site-by-site approach to assessing risk is needed, which takes into account the individual environmental characteristics of soils and human activities.

References

Apitz, S.E. (2008). Is risk based, sustainable sediment management consistent with European policy. Journal of Soils and Sediments 8: 461-466.

Ambast, S.K. and Sen, H.S. (**2006**). Integrated water management strategies for coastal ecosystem.*J. Indian Soc. Coastal Agric. Res. 24* :23–29.

Ashraf, M. A., Maah, M. J. and Yusoff, I. (2014).Soil Contamination, Risk Assessment and Remediation. pp 1-54.

Bhattacharyya , R., Ghosh, B. N., Mishra, P. K.,Mandal, B., Rao, C. S., Sarkar, D., Das, K., Anil, K. S., Lalitha, M., Hati, K. M. and Franzluebbers, A. J. (2015). Soil Degradation in India: Challenges and Potential Solutions. *Sustainability.7* :3528-3570.

Bhattacharyya, R., Kundu, S., Srivastva, A. K., Gupta, H. S., Prakash, V. and Bhatt, J.C. (**2011**). Long termfertilization effects on soil organic carbon pools in a sandy loam soil of the Indian Himalayas.*Plant Soil, 341* : 109–124.

Bhat, J.A., Kundu, M.C., Hazra, G.C. and Santra, G.H. (2010).Mandal, B. Rehabilitating acid soils for increasing crop productivity through low-cost liming material. *Sci. Total Environ.* **312:224-239.**

Commission Proposal COM (2006) 232 of 22 September 2006 for a Directive of the European Parliament and of the Council establishing a framework for the protection of soil and amending Directive 2004/35/EC.

Dhruvanarayan, V. V. N. and Ram, B. (1983). Estimation of soil erosion in India.*J. Irrig. Drain. Eng. 109*:419–434.

Evans, J., Wood, G. and Miller, A. (2006). The risk assessment-policy gap: An example from the UK contaminated land regime. Environment International 32: 1066-1071.

FosuMensah, B.Y., Okoffo, E. D.,Darko, G. and Gordon, C. (2016). Assessment of organochlorine pesticide residues in soils and drinking water sources from cocoa farms in Ghana. *Springer Plus*. 5: 869, 1-13.

Ghosh, K., Das, I., Saha, S., Banik, G. C., Ghosh, S., Maji, N.C. and Sanyal, S.K. (2004). Arsenic chemistry in groundwater in the Bengal Delta Plain: Implications in agricultural system. *J. Indian Chem.Soc.81*: 1–10.

Huinink, J.T.M. (1998). Soil quality requirements of use in urban environments. *Soil and Tillage Research*. 47: 157-162.

Kumar, B., Gaur, R., Goel, G., Mishra, M., Singh, S.K., Prakash, D., Kumar, S. and Sharma, C.S. (2011). Residues of Pesticides and Herbicides in Soils from Agriculture Areas of Delhi Region, India. *Journal of Environment and Earth Science*. 1 (2) :1-8.

Leon Paumen, M. (2008). Invertebrate life cycle responses to PAC exposure.PhD thesis. Amsterdam: University of Amsterdam.

Murugan, A.V.,Swarnam, T.P. and Gnanasambandan, S. (2013). Status and effect of pesticide residues in soils under different land uses of Andaman Islands, India. *Environ. Monit. Assess*. 185(10): 8135-45.

Mohan, A. and Sajayan, J. (2015). Soil pollution-A Momentous Crisis. *International Journal of Herbal Medicine*. 3(1): 45-47

Nathanail, P., McCaffrey, C., Earl, N., Forster, N.D., Gillett, A.G. and Ogden, R. (2005).A deterministic method for deriving site-specific human health assessment criteria for contaminants in soil. Human and Ecological Risk Assessment 11: 389-410.

National Bureau of Soil Survey & Land Use Planning (NBSS &LUP). *Soil Map (1:1 Million Scale)*; NBSS&LUP: Nagpur, India, 2004.

Okrent D. (1999). On intergenerational equity and its clash with intergenerational equity and on the need for policies to guide the regulation of disposal of wastes and other activities posing very long time risks. Risk Analysis 19: 877-901.

Provoost, J., Cornelis, C. and Swartjes, F. (2006). Comparison of soil clean-up standards fort race elements between countries: why do they differ? Journal of Soil and Sediments 6: 173-181.

Patterson, M.M., Cohen, E,., Promniei, H., Thomas, D.G., Rhodes, S., McKinley, A l. (2007). Origin of mixed brominated ethane groundwater plume: contaminant degradation pathways and reactions. Environmental Science & Technology 41: 1352-138.

Rao, G.V.R.,Kumari, B.R., Sahrawat, K.L. and Wani S.P. (2015). Integrated Pest Management (IPM) for Reducing Pesticide Residues in Crops and Natural

Resources.A. K. Chakravarthy (ed.), *New Horizons in Insect Science: Towards Sustainable Pest Management*.397-412.

Reddy, V.R. (**2003**). Land degradation in India: Extent, costs and determinants. *Econ. Polit. Wkly.* 38, 4700–4713.

Rombke, J. (2006). Tools and techniques for the assessment of eco toxicological impacts of contaminants in the terrestrial environment. *Human and Ecological Risk Assessment* 12: 84-101.

Richardson, G.M., Bright, D.A. and Dodd, M. (2006). Do current standards of practice in Canada measure what is relevant to human exposure at contaminated sites? II: oral bioaccessibility of contaminants in soil. *Human and Ecological Risk Assessment,* 12: 606-618.

Srinivasarao, C.H., Venkateswarlu, B., Lal, R., Singh, A.K. and Sumanta, K. (2013). Sustainable management of soils of dry land ecosystems for enhancing agronomic productivity and sequestering carbon.*Adv. Agron..121,* 253–329.

Singh, G., Ram, B., Narain, P., Bhushan, L.S. and Abrol, I.P.(1992). Soil erosion rates in India. *Ind. J. Soil Conserv.* 47, 97–99.

Sharma, N.K., Ghosh, B.N., Khola, O.P.S. and Dubey, R.K. (2013). Residue and tillage management for soil moisture conservation in post maize harvesting period under rainfed conditions of north-west Himalayas. *Ind. J. Soil Conserv.42,* 120–125.

Swartjes, F.A. (1999). Risk-based assessment of soil and groundwater quality in the Netherlands: standards and remediation urgency. Risk Analysis 19:1235-1248.

Sahu, H.B. and Dash, S. (2011). Land degradation due to Mining in India and its mitigation measures.In Proceedings of the Second International Conference on Environmental Science and Technology, Singapore.

Tarazona, J.V., Fernandez, M.D. and Vega, M.M. (2005). Regulation of contaminated soils in Spain. *Journal of Soil and Sediments* 5:121-124.

Urzelai, A., Vega, M. and Angulo, E. (2000). Deriving ecological risk-based soil quality values in the Basque Country. Science of the Total Environment 247: 279-284.

Van Zorge, J. A. (1996). Exposure to mixtures of chemical substances: is there a need for regulations. Food and Chemical Toxicology 34, 1033-1036.

Yadav, I.C., Syed, J.H., Cheng, Z., Zhang, G. and Jones, K.C. (2015). Current status of persistent organic pesticides residues in air, water, and soil, and their possible effect on neighboring countries: a comprehensive review of India. *Sci. Total Environ.* 123-37.

Study of Genetic Variability, Correlation and Path Analysis in Finger Millet (*Eleusine coracana* Gaertn)

D.D.Kadam

Officer In charge, Agricultural research Mohol, Mahatma Phule Agricultural University, Rhuri Solapur-413213(m.s)

India

Abstract

Wide range of variability was present in the elite line of Finger Millet under study PCV and GCV were high for plant height; days to 50 per cent flowering, flag leaf blade length, inflorescence length and yield. Heritability and genetic advance was high for leaf blade length, basal tiller and plant height. There were highly significant positive association with yield and almost all the growth and yield contributing characters except flag leaf blade width and exertion. Path analysis showed that indirect effect of yield had masked the direct or indirect effect in almost all the characters except inflorescence width. However the direct effect was in negative direction for characters viz; plant height, exertion panicle branch number inflorescence width length of largest finger resulted in non significant correlation with yield. Therefore yield, flag leaf blade length, flag leaf sheath length, inflorescence length, inflorescence width and 1000 grain weight must be given due emphasis in selection programme of genotypes for substantial high yield improvement of finger millet.

Keywords : Eleusine coracana variability, co-relation and path analysis

Finger Millet (*Eleusine coracana* Gaertn) locally known as Nachani/ Nagali/ Ragi/ Marua is an important food crop of tribal people/ below poverty line people. It is a rich source of minerals like calcium and iron and has biological efficiency in human nutrition and very much preferred by people who perform hard daily manual work, because of its good sustaining value. The success in plant breeding research is closely linked to variability in appropriate germ-plasm. Finger Millet is self pollinated crop and posses problem for the breeder's for exploiting the available genetic variability for selecting the promising genotypes.

Contact crossing method and pure line selection is the main method for finger millet improvement and development of new varieties. For this, a better understanding of the characters showing variability along with their genetic advance and heritability and their interactions for obtaining a high yielding genotypes is most important. Besides this, the knowledge of association of yield and yield contributing characters, directly as well as indirectly helps in achieving success in a plant breeding programme. Therefore, an experiment was conducted to determine the variability present, interrelationship among the yield traits and direct and indirect association among them through path co-efficient analysis in finger millet.

Materials and methods

Sixty five genotypes received from ICRISAT, Hydrabad with five checks were grown in three replications in randomized block design at All India co-ordinated Small Millets Improvement Project, Zonal Agril. Research Station, Shenda Park, Kolhapur. Each entry was grown on one meter length row with spacing of 22.5 cm between the rows and 10 cm within two plants. All the recommended package of practices were followed. Five plants randomly selected from each genotype in each replication were used to record observation on days to 50 per cent flowering [FLG], plant height [PLHT] (cm), basal tillers [BT], flag leaf blade length (cm) [FLBL], flag leaf blade width (cm) [FLBW], flag leaf sheath length (cm) [FLSL], peduncle length (cm) [PEDELEN], exertion (cm) [EXER], inflorescence length (cm) [INFLL], Inflorescence width (cm) [INFLW], length of longest finger (cm) [LLF], width of longest finger (cm) [WFL], panicle branch number [PBN], 1000 grain weight (gram) and grain yield $^{-1}$ plot [YIELD]. The mean of five plants was subjected to statistical analysis; data were statistically analysed to estimate phenotypic and genotypic co-efficient of variation as suggested by Burton (1952); phenotypic and genotypic co-relation by Panse and Sukhatme (1961) and heritability by Allard (1960) along with path co-efficient as suggested by Dewey and Lu(1959).

Results and discusions

Analysis of variance showed that genotypes differed significantly among themselves for all traits studied indicating presence of wide range of variability in the genotypes, Days to 50 per cent flowering ranged from 70 days to 96 days, plant height ranged from 61 cm to 106 cm, basal tillers ranged form 2-6. There was also wide range in yield and yield contributing characters. The range between 1000 grain weight and yield^{-1} plot was 1.21-2.81 and 26.0-542.7 gram^{-1} plot (Table 1). Phenotypic and genotypic co-efficient of variation given in Table 1 indicated that, there were no much differences among themselves indicating little effect of environment on their expression. GCV and PCV were high for 1000 grain weight and yield. Heritability was also very high for yield (98%), plant height (91%), flag leaf sheath length (87%), longest finger (84%), basal tiller (82%) and exertion (80%). There was fairly high heritability for other traits which ranged between 69 to 77 percent except exertion and panicle branch number. These results are in conformity with the studies made by Chaudhari and Acharya (1969), and Dhagat *et al.* (1972). High heritability was associated with high genetic gain in characters like plant height, flag leaf sheath length, and yield indicating that additive effects are important in determining these characters. These results are in agreement with studies of Kempana and Thirumalchar (1968). Genetic advance was high in longest finger, days to 50 percent flowering, exertion, inflorescence width, length of longest finger, width of longest finger, panicle number and 1000 grain weight, were quite less though high amount of heritability was present which attributed to lesser amount of variation also reflected by PCV and GCV for these characters. Characters having high GCV also have high genetic advance. Genetic advance was highest in case of yield, plant height, flag leaf blade length, flag leaf sheath length and days to 50 per cent flowering. Thus, selection for such characters will be useful for varietal improvement in finger millet. These results are in conformity with Goud and Laxmi (1977). A wide range of variability for all the characters associated with yield and high heritability which is predominantly due to additive genes reported in the studies made so far suggests greater scope for developing varieties superior to presently studied genotypes through hybridization followed by selection. Selection for 1000 grain weight will be more efficient as it is based on one or more highly heritable characters correlated with each other, which can be studied with the help of genotypic and phenotypic correlations. The genotypic correlation was lower than phenotypic ones, which indicated that environment played role in the inherent relationship between these characters which was similar to Johnson *et al.* (1955) 1000 grain weight and total yield was highly correlated with at 1% level of significance with all the characters under study in positive direction except for the exertion (Table 2). These results are also in agreement with Chaudhari and Acharya (1969) who reported the yield was highly positively and significantly associated with 1000 grain yield. The major variables of yield seem to be number of productive tiller and 1000 grain weight with inflorescence length.

Table 1: Mean , Range. Variance Co efficient Heritability and Genetic advance for different characters in finger millet.

Characters	Mean	Range	GCV	PCV	Heritability	Genetic advance	Genetic advance as percentage of mean % mean
FLG	80.16	70-96	83.40	94.67	77.60	12.44	15.13
PLHT	80.91	61-106	45.64	52.76	90.88	23.14	28.59
BT	4.57	2-6	72.68	40.65	81.95	23.69	18.61
FLBL	33.15	15.41	75.36	20.02	76.73	30.44	31.65
FLBW	0.84	0.53-1.26	51.02	19.75	58.55	0.20	23.81
FLSL	10.95	7.0-25.33	25.44	27.32	86.69	5.34	48.79
PEDELEN	21.47	6.0-28.60	20.52	24.43	72.91	7.75	36.11
EXER	10.98	5.0-14.66	13.63	31.35	8.10	5.75	52.36
INFLL	6.50	4.53-5.33	21.59	25.93	69.30	2.43	14.37

Characters	Mean	Range	GCV	PCV	Heritability	Genetic advance	Genetic advance as percentage of mean % mean
INFLW	4.43	2.36-7.00	15.81	23.67	44.62	0.96	21.76
LLF	5.89	3.66-16.33	72.52	35.54	83.76	3.61	61.52
WLF	0.77	0.57-1.03	17.39	26.22	43.99	0.18	23.76
PBN	1.22	1.00-3.00	25.01	40.03	39.04	0.39	32.19
1000 Grain Weight	2.20	1.21-2.81	87.14	83.16	78.10	0.60	27.07
YIELD	190.48	26.0-542.7	88.59	89.19	98.00	97.59	94.8

Studies made by Dhagat *et al.* (1978) are in line for various characters who had positive and significant correlation of plant height with peduncle length, inflorescence width, flag leaf sheath length and exertion with plant height was reported. Similarly, path coefficient under study analysis of selected characters indicated that yield is having direct correlation with No. basal tiller, 1000 grain weight flag leaf length, flag leaf width, inflorescence length inflorescence width. The major variables of yield seem to be basal tiller, inflorescence length, inflorescence width, and days to 50 per cent flowering.

Selection based on correlation may be misleading. Hence, path analysis for 1000 grain weight was studied (Table 3). The direct effects of panicle branch number was negative direction with though the correlation was significant and in positive direction. Though direct effect of plant height, basal tiller 1000 grain weight, inflorescence length, inflorescence width and yield had high positive direct effect on 1000 grain weight indirect effect of yield with plant height, flag leaf length flag leaf width, flag leaf sheath length, flag leaf sheath width and 1000 grain weight and; exertion was very high resulting in positive significant correlation in these characters indirect effect on yield had masked the direct or indirect effect in almost all the characters except plant height where the indirect effect were in negative direction via flag leaf sheath length flag leaf sheath width exertion panicle branch number resulted in non significant correlations. Though the direct effect was high and in positive direction. These finding are similar to Dhagat *et al.* (1972) where positive direct as well as indirect effect via 1000 grain weight. The direct effect at inflorescence width yield was positive with low magnitude also reported by Kempana and Thirumalacher (1968). Direct effect of 1000 grain weight on yield was fairly high and positive; and its indirect effects via inflorescence length, which was in agreement with the studies made by Dhagat *et al.* (1972). Thus, the results indicate that one should emphasize on yield basal tiller, inflorescence length and Inflorescence width, and 1000 grain weight while making selection for improvement in the yield of finger millet.

Table 2. Phenotypic (above diagonal and genotypic below diagonal correlation coefficient for different pair of characteristics in finger millet.

Sr. No.	Character	1	2	3	4	5	6	7	8	9	10	11	12	13	14	15
1	FLG		0.230**	0.051**	0.251**	-0.012	0.063**	0.250**	0.042**	0.231**	0.358**	0.308**	0.077**	0.092**	0.116**	0.454**
2	PLHT	0.221		0.039	0.274**	0.122*	0.147**	0.056	-0.012	0.111*	0.098**	0.218**	0.166**	-0.230**	0.083**	0.418**
3	Basal tillers	-0.725	-0.873		0.036**	-0.037	0.185**	0.055**	-0.058*	0.073**	0.197**	0.058**	0.282**	0.061**	-0.162	0.603**
4	FLBL	1.89	22.455	-0.442		0.243**	0.025**	-0.092*	-0.127	0.156*	0.236**	0.218**	0.225**	0.149**	-0.010	0.603**
5	FLBW	-0.015	0.251	-0.011	0.268		0.088**	0.066**	-0.010	0.067**	0.111**	0.105**	0.215**	-0.045	0.307**	0.503**
6	FLSL	1.69	5.411	1.013	0.486	-0.043		-0.006	0.063**	0.098**	0.005	0.054**	-0.018	0.069**	-0.99**	0.547**
7	PEDLEN	-1.305	-3.597	0.511	3.164	0.0567	-0.099		0.508*	-0.98**	0.005	0.054**	-0.008	0.069**	0.100**	0.410**
8	Exer	-1.18	0.903	0.363	-2.896	0.0190	-0.690	-9.030		0.056**	0.116**	-0.016	-0.123*	-0.008	0.124**	0.050
9	INFLL	3.003	7.334	-0.228	1.763	0.019	-0.496	-1.132	0.329		0.300*	0.563**	0.049*	0.097**	0.042	0.091**
10	INFLW	2.913	1.262	-0.377	1.641	0.019	0.015	-0.580	-0.411	0.535		0.417*	0.195**	0.115**	0.069*	0.323**
11	LLF	5.023	5.646	-0.222	3.043	0.035	0.340	-1.194	-0.117	2.007	0.915		0.032**	0.081**	0.104**	0.415**
12	WLF	0.12	0.414	0.010	0.296	0.007	-0.005	-0.019	-0.186	-0.098	-0.083	-0.012		0.123**	0.177**	0.385**
13	PBN	-0.340	-1.388	0.655	0.481	-0.004	0.100	0.271	-0.113	-0.080	-0.059	-0.083	0.012		-0.044*	0.575**
14	1000 Gr. Wt. (gram.)	0.334	0.239	-0.110	-0.025	0.002	-0.115	-0.333	-0.158	0.026	0.027	0.080	0.013	0.080		0.785**
15	Grain plot yield (g)	11.33	-25.463	13.291	-19.858	1.926	-49.524	66.300	19.459	-17.573	-14.530	-30.179	2.106	4.793	15.14	

*, ** Significance at 5 and % level of significance, respectively

Table 3. Genotypic Path Co-efficient analysis showing and indirect important characters on finger millet

Character	1	2	3	4	5	6	7	8	9	10	11	12	13	14	Correlation
1 FLG	**46.945**	0.291	-0.449	0.292	0.008	0.066	-0.312	-0.075	0.313	0.526	0.361	0.102	-0.197	0.105	-0.449**
2 PLHT	23.553	**138.840**	-0.064	0.335	0.158	0.258	-0.086	-0.033	0.145	0.179	0.261	0.285	-0.336	-0.067	-0.150**
3 BT	-0.510	-1.243	**2.748**	-0.375	-0.106	0.261	-0.086	-0.055	-0.110	-0.314	-0.060	-0.050	0.119	-0.185	-0.203**
4 FLBL	11.630	22.976	-0.361	**33.791**	0.332	0.024	-0.171	-0.197	0.154	0.323	0.264	0.317	-0.327	-0.070	-0.443**
5 FLBW	0.006	0.237	-0.022	0.245	**0.016**	-0.174	0.091	0.058	0.044	0.137	0.118	0.365	-0.176	0.037	0.547**
6 FLSL	1.263	5.200	0.948	0.391	-0.061	**7.745**	-0.046	-0.104	-0.119	-0.043	0.063	0.005	0.075	-0.126	-0.210**
7 PEDLE	-9.422	-4.486	0.647	-4.386	-0.051	-0.565	**19.430**	0.558	-0.164	-237.00	-0.126	0.024	0.211	-0.213	0.060
8 EXER	-1.609	-1.210	-0.281	-3.550	-0.023	-0.898	7.583	**9.620**	0.632	-0.232	-0.028	-0.129	0.034	-0.167	0.130**
9 INFLL	3.039	2.421	-0.259	1.266	-0.008	-0.472	-1.026	0.276	**2.033**	0.424	0.640	0.080	-0.005	0.070	-0.345**
10 INFLW	2.522	1.479	-0.365	1.317	0.012	-0.084	-0.734	-0.503	0.420	**0.489**	0.542	0.293	-0.322	0.141	-0.321**
11 LLF	4.447	5.908	-0.189	2.946	0.286	0.333	-1.067	-0.165	1.733	0.777	**3.673**	-0.023	-0.076	0.109	-0.385**
12 WLF	0.093	0.450	-0.110	0.247	0.006	0.002	0.014	-0.054	0.015	0.028	-0.006	**0.018**	-0.305	0.334	0.384**
13 PBN	-0.411	-1.206	0.066	0.579	-0.007	0.063	0.284	0.032	-0.002	-0.069	-0.044	-0.013	**-0.093**	-0.073	0.182**
14 YIELD	113.951	-32.005	11.6676	20.128	2.147	48.612	60.234	15.452	-19.077	-14.600	31.564	2.299	6.204	14.994	0.785**

Residual effect =0.3394, Bold Letter s denote direct effect

** Significance at 1% level probability, * Correlation with 1000 grain weight

Acknowledgement

The authors express their gratitude to the University, ICAR and ICRISAT authority for their encouragement and providing the necessary facilities to carry out the present research work. This study was supported by an All India Co-ordinated Small Millet Improvement Project, Zonal Agriculture Research Station, Kolhapur Mahatma Phule agricultural university Rahuri,(Maharashtra).

References

Allard, R.W. (1960). Principles of Plant Breeding John Willey and Sons. Inc. New York Pp. :85-95.

Burton, G.M.(1952). Quantitative inheritance ingrasses proc, 6th Int, Grass land Cong 1:277-283

Chaudhari, L.B, and B.C. Acharya, 1969. Genotypic variability and path co-efficient analysis of component of ragi (*Eleusine caracana* Gaertn) Expt, agric, Sci.:295-300.

Dewey,D.R. and K.H.Lu.(1959). A co-relation and path co-efficient analysis of components of crested wheat grass seed production, Agron J.51; 515-518.

Dhagat, N.K., G.L. Pattider, R.S. Shrivastava and R.C. Joshi. (1972). Correlation Genetic variability study in ragi JNVKVV Res. J. 6:121-124.

Dhagat, N.K., U. Goswami and V.G. Narsinghani. (1978). Genetic variability, character correlation and path analysis of Barnyard Millet (*Echinochola crusgulli* (L)) Beauv Indian J. Agril Sci48:211-214.

Goud, J.V. and P.V. Laxmi., (1977). Morphological and genetical variability for quantitative characters in ragi (*Eleusine caracana* Gaertn) Mysore J. of Agril Sci. 11:438-443.

Johnson, H.W., H.F. Gobinson and R.E. Comstock. (1955). Estimate of genetic and environmental variability in soybean. Agron. J.47:314-318.

Kempana C and D.K.Thirumalchar (1968). Studies on the genotypes variation in ragi (*Eleusive coracana* Gaertn) Mysore J. Agric Sci. 2:29-34.

Panse, V.G. and P.V. Sukhatme. (1961). Statistical methods for agricultural workers ICAR, New Delhi.

Singh, R.K. and Chaudhari,. (1979. Biometrical Techniques in Genetics and Breeding. International Bioscience Publisher Hissar (India).

www.ingramcontent.com/pod-product-compliance
Lightning Source LLC
Chambersburg PA
CBHW072254210326
41458CB00073B/1707